Workbook to Accompany Maintenance Fundamentals for Wind Technicians

Wayne Kilcollins

Australia • Brazil • Japan • Korea • Mexico • Singapore • Spain • United Kingdom • United States

Workbook to Accompany Maintenance Fundamentals for Wind Technicians
Wayne Kilcollins

Vice President, Careers & Computing:
  Dave Garza

Director of Learning Solutions: Sandy Clark

Executive Editor: Dave Boelio

Associate Acquisitions Editor: Nicole Sgueglia

Director, Development-Career and Computing:
  Marah Bellegarde

Senior Product Manager: Sharon Chambliss

Editorial Assistant: Leah Costakis

Senior Brand Manager: Kristin McNary

Market Development Manager: Erin Brennan

Senior Production Director: Wendy Troeger

Production Manager: Mark Bernard

Senior Content Project Manager: Cheri Plasse

Art Director: GEX, Inc.

Media Editor: Deborah Bordeaux

© 2014 Delmar, Cengage Learning

ALL RIGHTS RESERVED. No part of this work covered by the copyright herein may be reproduced, transmitted, stored, or used in any form or by any means graphic, electronic, or mechanical, including but not limited to photocopying, recording, scanning, digitizing, taping, Web distribution, information networks, or information storage and retrieval systems, except as permitted under Section 107 or 108 of the 1976 United States Copyright Act, without the prior written permission of the publisher.

For product information and technology assistance, contact us at
Cengage Learning Customer & Sales Support, 1-800-354-9706
For permission to use material from this text or product,
submit all requests online at **www.cengage.com/permissions**
Further permissions questions can be e-mailed to
permissionrequest@cengage.com

Library of Congress Control Number: 2011944382

ISBN-13: 978-1-111-30775-2

ISBN-10: 1-111-30775-X

Delmar
5 Maxwell Drive
Clifton Park, NY 12065-2919
USA

Cengage Learning is a leading provider of customized learning solutions with office locations around the globe, including Singapore, the United Kingdom, Australia, Mexico, Brazil, and Japan. Locate your local office at
www.cengage.com/global

Cengage Learning products are represented in Canada by Nelson Education, Ltd.

To learn more about Delmar, visit **www.cengage.com/delmar**

Purchase any of our products at your local college store or at our preferred online store **www.cengagebrain.com**

Notice to the Reader
Publisher does not warrant or guarantee any of the products described herein or perform any independent analysis in connection with any of the product information contained herein. Publisher does not assume, and expressly disclaims, any obligation to obtain and include information other than that provided to it by the manufacturer. The reader is expressly warned to consider and adopt all safety precautions that might be indicated by the activities described herein and to avoid all potential hazards. By following the instructions contained herein, the reader willingly assumes all risks in connection with such instructions. The publisher makes no representations or warranties of any kind, including but not limited to, the warranties of fitness for particular purpose or merchantability, nor are any such representations implied with respect to the material set forth herein, and the publisher takes no responsibility with respect to such material. The publisher shall not be liable for any special, consequential, or exemplary damages resulting, in whole or part, from the readers' use of, or reliance upon, this material.

Printed at CLDPC, USA, 12-18

Table of Contents

Chapter 1 – Introduction 1

Chapter 2 – Tower Safety 21

Chapter 3 – Workplace Safety 29

Chapter 4 – Lubrication 37

Chapter 5 – Fluid Power 45

Chapter 6 – Bolting Practices 55

Chapter 7 – Test Equipment 63

Chapter 8 – Component Alignment 73

Chapter 9 – Down Tower Assembly 83

Chapter 10 – Tower 93

Chapter 11 – Machine Head 103

Chapter 12 – Drive Train 115

Chapter 13 – Generator 127

Chapter 14 – Rotor Assembly 137

Chapter 15 – External Surfaces 145

Chapter 16 – Developing a Preventative Maintenance Program 151

Chapter 17 – Wind-Farm Management Tools 163

Appendix A – Hydraulic Symbols 171

Appendix B – Electrical Symbols 177

Appendix C – Bolting Notes 181

Appendix D – Useful Information 183

CHAPTER 1

Introduction

Name _____   Date _____

Determine the surface area for each of the following shapes:

1. Area = _____

2. Area = _____

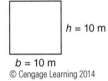

3. Area = _____

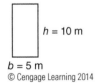

4. Area = _____

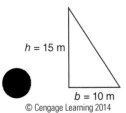

5. Area = _____

Worksheet 1–1

2 Using Resources

Name _____   Date _____

Determine the surface area for each of the following shapes. *Note*: Break the shapes into multiple sections, determine the area for each section, and add the areas to find the total.

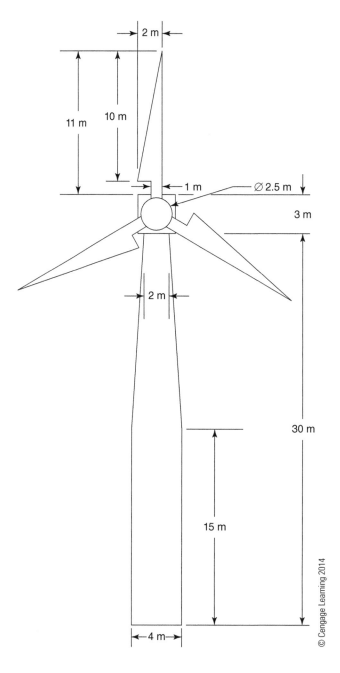

Blade area (3X): _____

Hub area: _____

Tower area: _____

Total area: _____

Worksheet 1–2

Name _____   Date _____

Determine instantaneous power for the following wind-turbine rotor assemblies using provided information and power (P_I) equation from the textbook. First, determine the rotor-assembly area that would be exposed to the oncoming wind. This area is considered the swept area (A) for the rotor assembly.

1. Horizontal axis wind turbine (HAWT): blade length (r) = 25 meters, rotor efficiency (Cp) = 45%, wind speed = 6 m/s. Use air density (ρ) at 20°C and mean sea level (MSL) for the calculation.

2. HAWT: blade length (r) = 50 meters, rotor efficiency (Cp) = 45%, wind speed = 6 m/s. Use same air density (ρ) as problem 1.

3. HAWT: blade length (r) = 25 meters, rotor efficiency (Cp) = 45%, wind speed = 12 m/s. Use same air density (ρ) as problem 1.

4. HAWT: blade length (r) = 50 meters, rotor efficiency (Cp) = 45%, wind speed = 12 m/s. Use same air density (ρ) as problem 1.

5. Compare the solutions and indicate which variable had the largest impact on the final results: swept area (A) or wind speed (v)? Why?

    _____

    _____

    _____

Worksheet 1–3

4   Using Resources

Name _____   Date _____

Determine the average annual wind speed for the area listed on the following map.

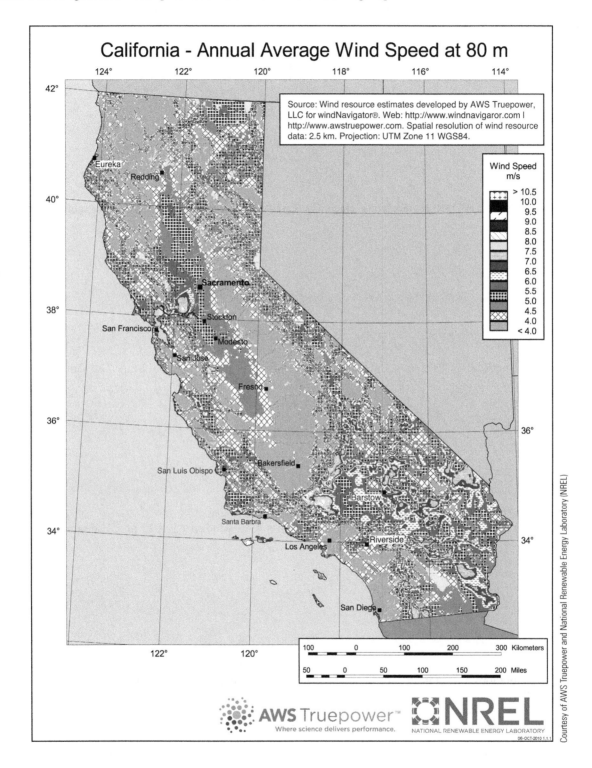

Area 1: N35°15′ W115°. Average annual wind speed at 80 m: _____

Worksheet 1-4

Name _____   Date _____

● Determine the average annual wind speed for the areas listed on the following map.

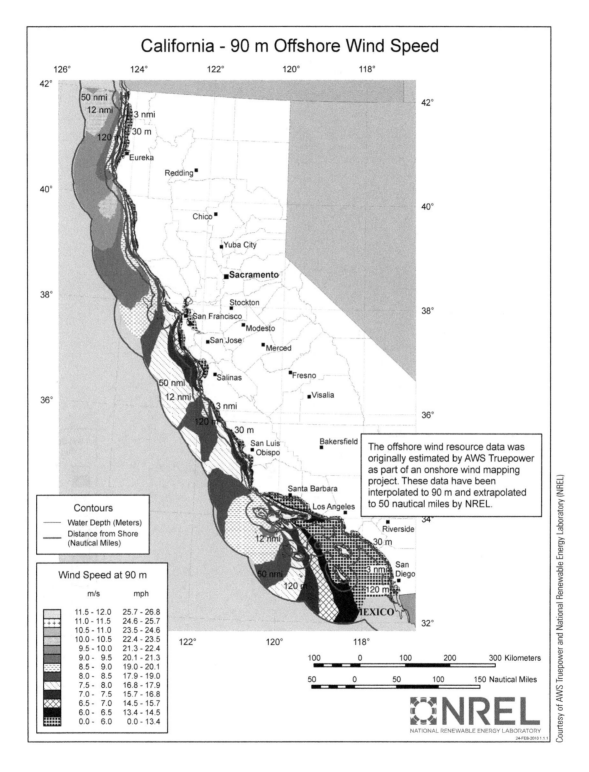

● Area 2: N34° W121°. Average annual wind speed at 90 m: _____

Area 3: N39°50′ W124°30′. Average annual wind speed at 90m: _____

6 Using Resources

Name _____ Date _____

Determine the average annual wind speed for the areas listed on the following map.

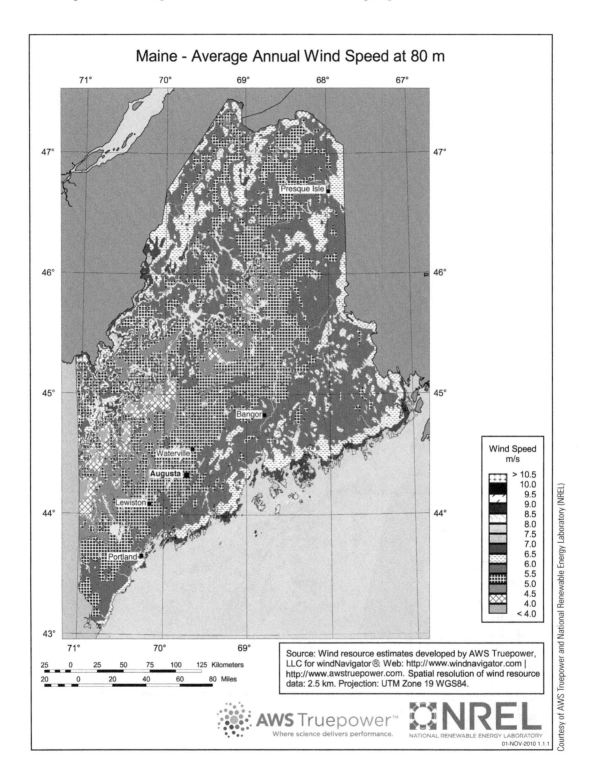

Area 1: N45°30′ W70°. Average annual wind speed at 80 m: _____

Area 2: N46°45 W67°50. Average annual wind speed at 80 m: _____

Chapter 1   Introduction   7

Name _____   Date _____

● Determine the average annual wind speed for the area listed on the following map.

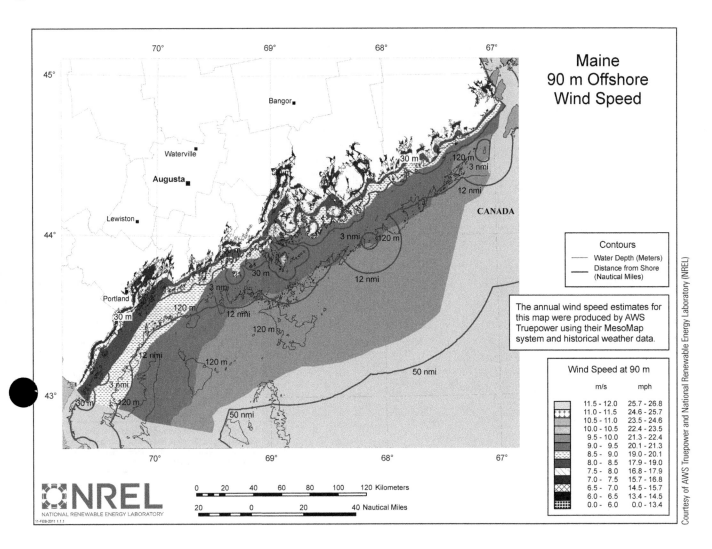

Area 3: N43°30′ W68°30′. Average annual wind speed at 90 m: _____

Worksheet 1-7

8   Using Resources

Name _____   Date _____

Determine the average annual wind speed for the areas listed on the following map.

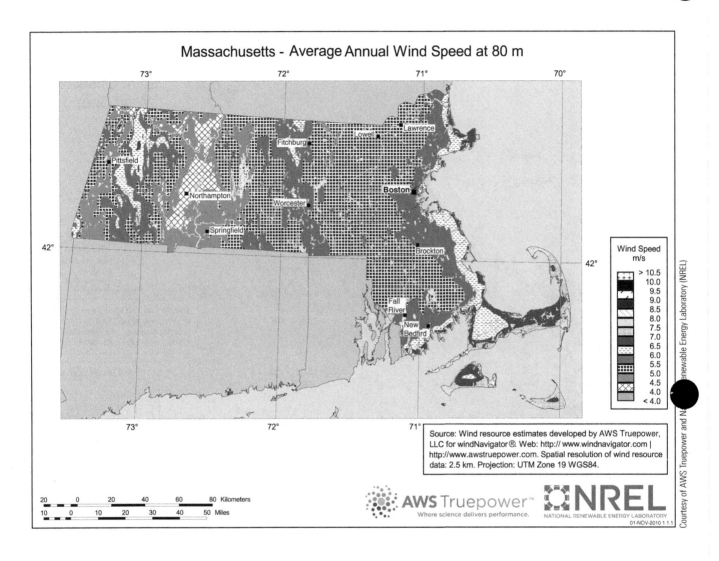

Area 1: N42°40′ W73°15′. Average annual wind speed at 80 m: _____

Area 2: N42°40′ W70°40′. Average annual wind speed at 80 m: _____

Worksheet 1–8

Chapter 1 Introduction 9

Name _____ Date _____

● Determine the average annual wind speed for the area listed on the following map.

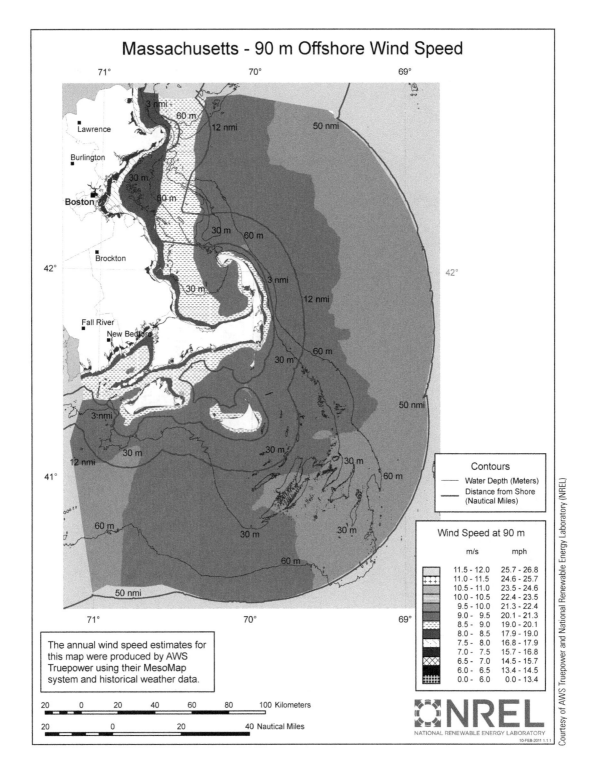

● Area 3: N42° W69°. Average annual wind speed at 90 m: _____

Worksheet 1–9

10   Using Resources

Name _____   Date _____

Determine the average annual wind speed for the area listed on the following map.

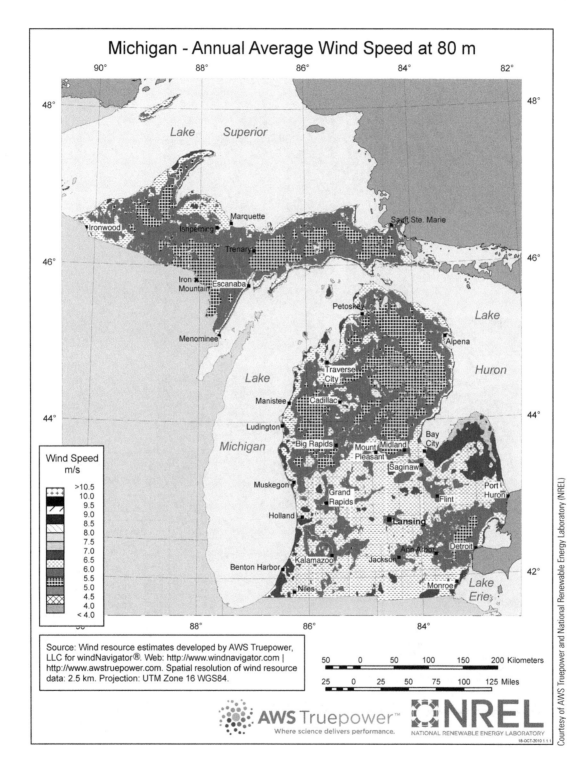

Area 1: N44° W83°: Average annual wind speed at 80 m: _____

Name _____ Date _____

- Determine the average annual wind speed for the areas listed on the following map.

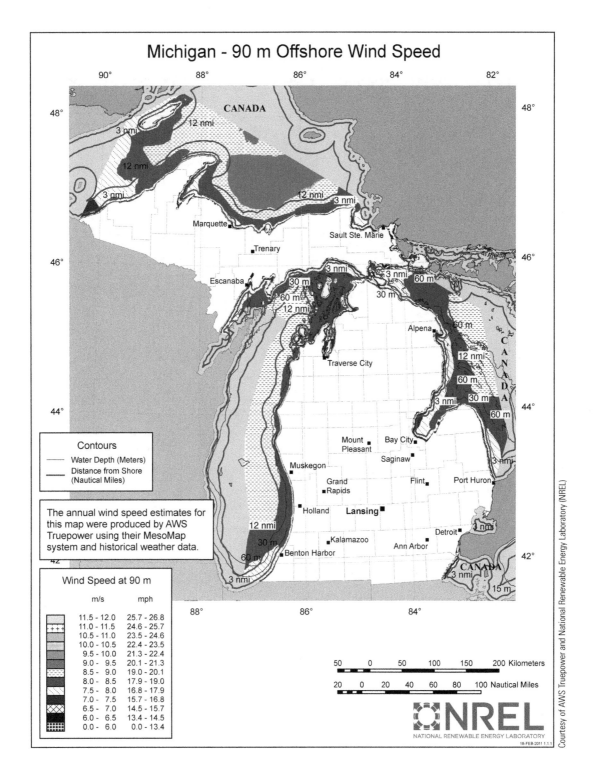

- Area 2: N43° W87°. Average annual wind speed at 90 m: _____

Area 3: N47° W87°. Average annual wind speed at 90 m: _____

Worksheet 1–11

Name _____   Date _____

Determine the average annual wind speed for the areas listed on the following map.

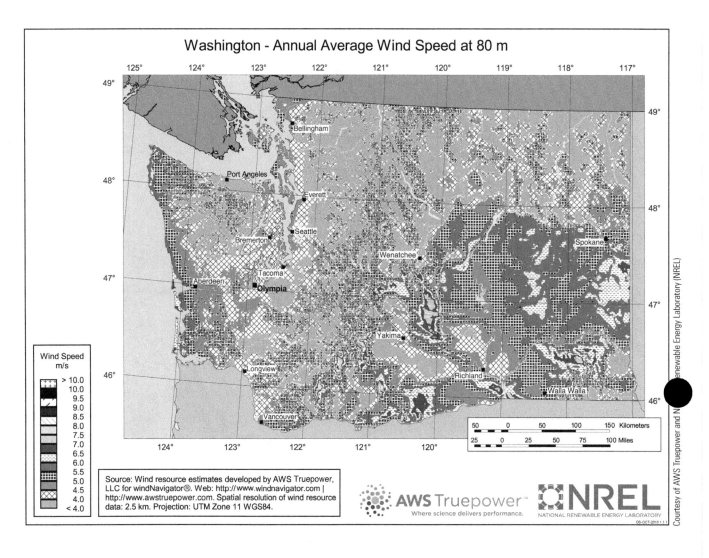

Area 1: N47° W120°10′. Average annual wind speed at 80 m: _____

Area 2: N48° W124°30′. Average annual wind speed at 80 m: _____

Name _____   Date _____

- Determine the average annual wind speed for the area listed on the following map.

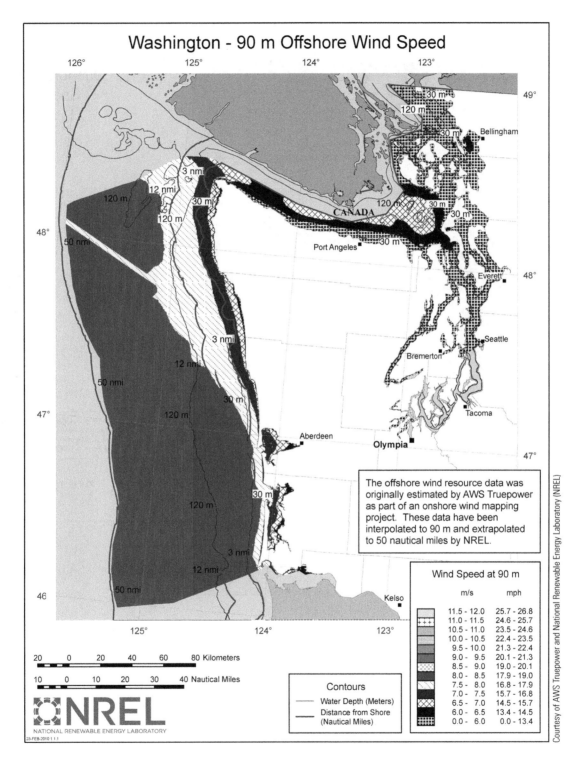

- Area 3: N47° W125°. Average annual wind speed at 90 m: _____

14 Using Resources

Name _____ Date _____

Determine the average annual wind speed for the areas listed on the following map.

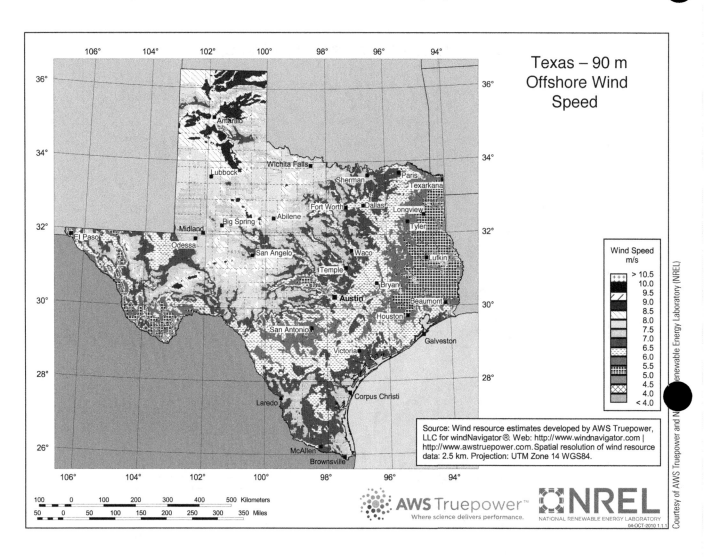

Area 1: N36°10′ W101°. Average annual wind speed at 80 m: _____

Area 2: N32°30′ W102°10′. Average annual wind speed at 80 m: _____

Worksheet 1–14

Name _____   Date _____

Determine the average annual wind speed for the area listed on the following map.

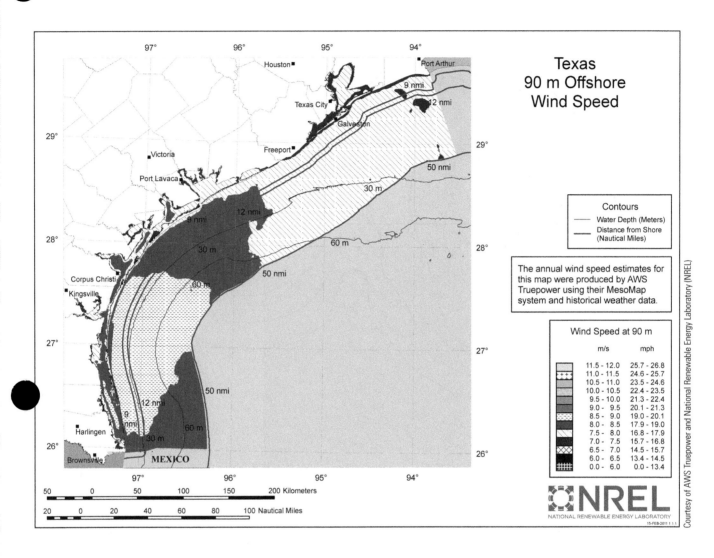

Area 3: N27° W97°. Average annual wind speed at 90 m: _____

16 Using Resources

Name _____ Date _____

Sketch the elevation view for terrain shown on the following topographical map.

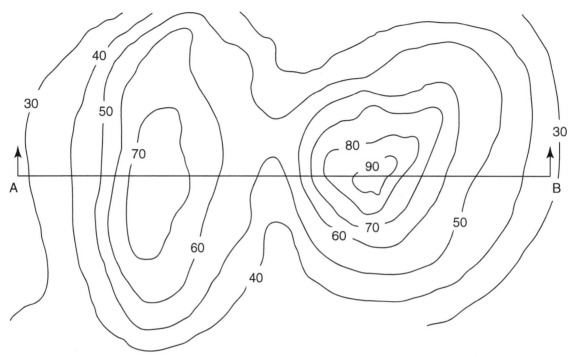

TOPOGRAPHIC MAP

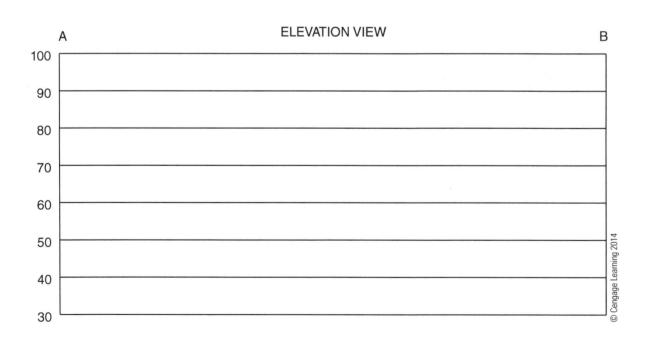

ELEVATION VIEW

Worksheet 1–16

Name _____ Date _____

Determine the distance between points A–B and B–C along with terrain slope for each line segment on the following topographic map.

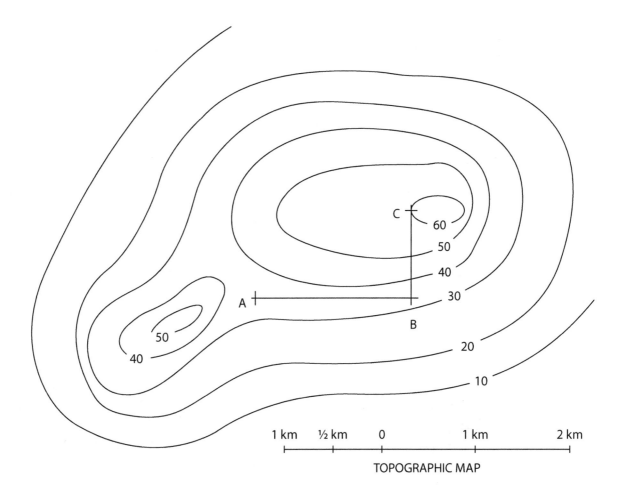

TOPOGRAPHIC MAP

70 m ─────────────────────────────────
60 m ─────────────────────────────────
50 m ─────────────────────────────────
40 m ─────────────────────────────────
30 m ─────────────────────────────────
20 m ─────────────────────────────────
10 m ─────────────────────────────────
 0 m ─────────────────────────────────

Worksheet 1–17

18 Using Resources

Name _____ Date _____

Match the topographic map view shown on the left to its corresponding elevation view on the right.

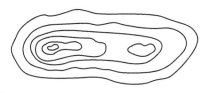

1

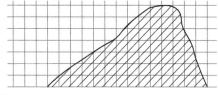

A

2

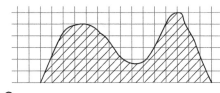

B

3

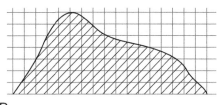

C

4

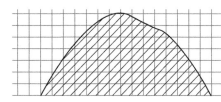

D

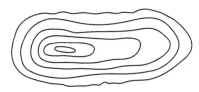

5

E

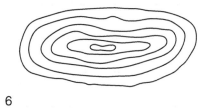

6

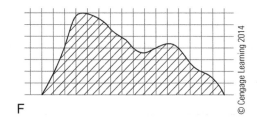

F

Worksheet 1–18

Name _____   Date _____

● This exercise shows the relationship between sound-power level and sound-pressure level for the following scenarios.

1. Determine the sound-power level (dB) of a freight train passing a residence at a distance of 500 meters away if the sound-pressure level is measured at 50 dBA outside the residence.

2. Determine the sound-pressure level (dBA) a ramp employee may be exposed to if an aircraft pilot is running up an engine during a preflight check. Estimated sound-power level of the engine is 110 dB, and the employee is standing 30 meters from the aircraft.

●

3. Determine the sound-pressure level (dBA) outside a residence located 5,000 meters from a wind turbine if the sound-power level is 95 dB at the wind-turbine nacelle.

4. Determine the sound-power level (dB) of a wind turbine (100-meter tower height) located on a 500-meter ridge at a horizontal distance 2,500 meters from a residence if the measured sound pressure outside the residence is 35 dBA. First calculate the straight-line distance between the wind-turbine nacelle and the residence located below the ridge.

●

20 Using Resources

Name _____ Date _____

This exercise shows the interaction of sound with solid objects. Set up a large area with a sound source located in the center and a couple of large solid objects at varying distances away from the source. Use a sound-measuring device to map the sound-pressure level at several distances from the source. Plot the sound-pressure levels on a polar coordinate chart as shown in the following and note any variations resulting from the objects. Why may the sound-pressure level be different behind an object compared to a direct line of sight to the source?

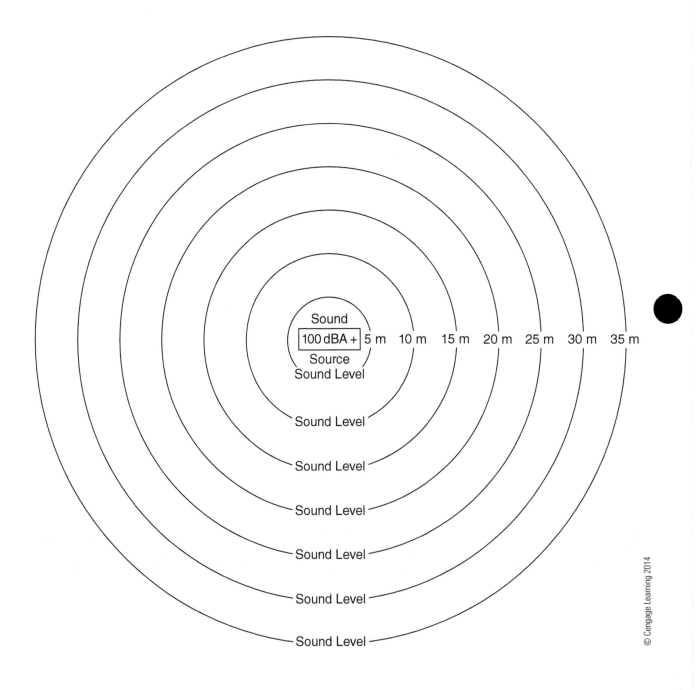

CHAPTER 2

Tower Safety

Name _____ Date _____

Choose the three methods used to prevent hazards in the workplace and describe the difference between each of these three methods:

Shotgun approach	Engineering	Hazard-risk category
Personal protective equipment	Overhead hazard zone	NFPA 70E
Anchorage	Management	OSHA
Fall prevention	Guardrails	

1. _____
_____
_____
_____
_____

2. _____
_____
_____
_____
_____
_____
_____

3. _____
_____
_____
_____
_____

Worksheet 2–1

Working at Heights

Name _____ Date _____

Match each of the following hazards to an engineering change that can eliminate or reduce the issue. *Note:* Some hazards may be eliminated by multiple changes, but some may require other methods.

Hazard	Engineering Change
Chemical burn	Overhead barrier
Electrocution	Guardrails
Fall	Ventilation
Falling object	Move activity to lower level
Heat exhaustion	Ground-fault circuit interrupter (GFCI)
Thermal burn	Interlock switch and full enclosure
Rotating shaft	Ladder cage
Hypothermia	Train employees to use caution when working around blind areas in remote locations such as culverts, steps, and cabinets
Pinch point	Reduce system voltage
Entanglement	Move activity to climate-controlled area
Flying debris	Safety glasses
Arc flash burn	Require qualified employee to perform activity
Asphyxiation	Barrier
Injection	Exposure suit
Snake bite	Warning lines

Worksheet 2–2

Name _____   Date _____

Match each of the following hazards to a management practice that may eliminate or reduce the issue. *Note:* Some hazards may be eliminated by multiple practices, but some may require other methods.

Hazard	**Management Practice**
Chemical burn	Do not get within 1,000 feet of a wind turbine if conditions can create ice falling from exterior surfaces.
Electrocution	Use personal flotation device.
Fall	Only water-survival–qualified employees may work on offshore platforms.
Falling object	Develop confined space policy, procedures, and training.
Heat exhaustion	Establish procedure to define approach boundaries around live electrical activities.
Thermal burn	Require that tools be connected with a lanyard during work on elevated platforms.
Rotating shaft	Exposure suit must be worn during work activity.
Pinch point	Implement procedure to reduce exposure time for elevated working temperatures.
Entanglement	Use safety glasses.
Flying debris	Only ANSI or CSA approved equipment to be used during activity at heights.
Arc flash burn	Require qualified employee perform activity.
Asphyxiation	Do not wear loose clothing around moving shafts.

Worksheet 2–3

24 Working at Heights

Name _____ Date _____

Match the following hazards with the appropriate personal protective equipment (PPE) used to eliminate or reduce the issue.
Note: Some hazards may be eliminated by multiple PPE options, but some may require other methods.

Hazard	**Personal Protective Equipment**
Chemical burn	Earplugs
Electrocution	Acid-resistant apron
Fall	Insulated dielectric gloves
Falling object	HRC-rated garments
Heat exhaustion	Insulated sleeves
Thermal burn	Hardhat
Rotating shaft	Goggles
Hypothermia	Full-body harness with shock-absorbing lanyards
Pinch point	Safety shoes
Entanglement	Limit time exposed to elevated temperatures and stay hydrated
Injection	Safety glasses
Arc flash burn	Require qualified employee to perform activity
Asphyxiation	Heat-resistant gloves
Flying debris	Full enclosure with interlock switch
Elevated sound level	Extreme temperature outerwear
Puncture wound	SCBA

Worksheet 2–4

Chapter 2   Tower Safety   **25**

Name _____   Date _____

Write below each of the following assemblies or components with their respective safety system function (fall-prevention or fall arrest):

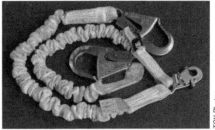

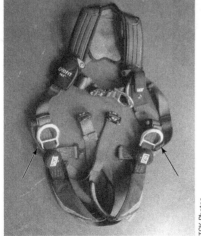

Worksheet 2–5   Page 1 of 2

26   Working at Heights

Name _____   Date _____

List the purpose of each of the following safety items as body support, connector, anchorage, or rescue system. *Note:* A connector is a device used to attach a body support system to an anchorage. This may include a device with an integral shock-absorbing device.

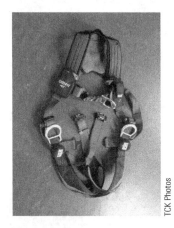

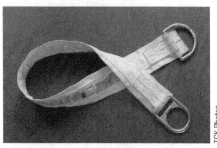

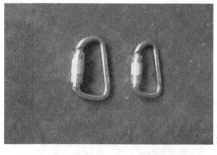

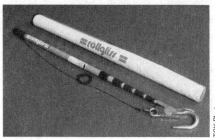

Worksheet 2–6

Chapter 2 Tower Safety **27**

Name _____ Date _____

List the defect for each of the following images and describe why the component should not be used for a fall-arrest system.

_____
_____
_____
_____

_____
_____
_____
_____

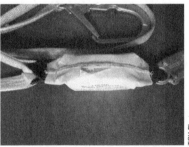

_____
_____
_____
_____

_____
_____
_____
_____

_____
_____
_____
_____

Worksheet 2-7   Page 1 of 2

28 Working at Heights

Name _____ Date _____

List the defect for each of the following images and describe why the component should not be used for a fall-arrest system.

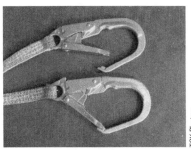

_____
_____
_____
_____

_____
_____
_____
_____

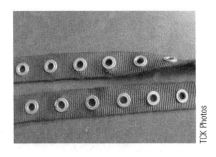

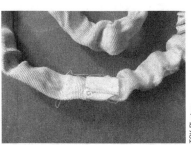

_____
_____
_____
_____

_____
_____
_____
_____

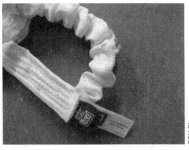

_____
_____
_____
_____

Worksheet 2-8

CHAPTER 3

Workplace Safety

Name _____ Date _____

Match the following work activity with an associated hazard that may be encountered. Some hazards may be associated with more than one activity. Choose one hazard that is the best match for the work activity.

Work Activity   **Associated Hazard**

Fluid-power troubleshooting   Drowning

Electrical troubleshooting   Rattlesnake bite

Mechanical maintenance   Acid burn

Chemical handling   Entanglement

Welding   Asphyxia

Blade-interior inspection   Arc-flash burn

Offshore transportation   Injection injury

Working around a tower in a remote site   Lung damage

Worksheet 3–1

General Safety

Name _____ Date _____

Match the following personal protective equipment (PPE) with a hazard risk category (HRC) for an electrical activity. Some items may be necessary for several activities, so they may be matched with the general case "All Categories."

Personal Protective Equipment (PPE)	**Hazard Risk Category (HRC)**
40 Cal/cm² AR coveralls	All categories
8 Cal/cm² AR flash-suit jacket	
8 Cal/cm² AR balaclava hood	HRC 1
Safety glasses or goggles	
Hearing protection	HRC 2
4 Cal/cm² AR long pants	
Leather safety shoes	HRC 2*
25 Cal/cm² flash-suit hood	
Hard hat	HRC 3
40 Cal/cm² AR jacket	
8 Cal/cm² AR full-face shield	HRC 4

*NFPA 70E electrical standard

Worksheet 3–2

Name _____   Date _____

Match the following PPE insulating electrical glove tag color with an associated maximum usage voltage.

Recommended Maximum Usage Voltage   **Insulating Glove Tag Color/Class**

500 VAC   Beige/00

11,250 VDC

36,000 VAC   Red/0

1,500 VDC

7,500 VAC   White/1

25,500 VDC

39,750 VDC   Yellow/2

26,500 VAC

750 VDC   Green/3

1,000 VAC

17,000 VAC   Orange/4

Worksheet 3–3

32 General Safety

Name _____ Date _____

Match the following tool with its common name.

Tool **Common Name**

Ratchet wrench

Combination wrench

Flat-blade screwdriver

Needle-nose pliers

Engineer's hammer

Wire strippers

Lineman pliers

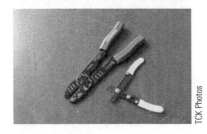

Phillips screwdriver

Worksheet 3–4

Name _____   Date _____

Match the following tool with its common name.

Tool   **Common Name**

   Portable drill

                                        Hydraulic torque wrench

                                        Hydraulic power unit

                                        Electric nut driver

                                        Torque wrench

                                        Hydraulic bolt tensioner

                                        Ratchet set

                                        Grease gun

Worksheet 3–5

34   General Safety

Name _____   Date _____

Match the following rigging equipment with its common name.

Rigging Equipment

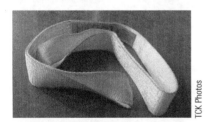

Common Name

Chain hoist

Synthetic web sling

Chain sling

Wire rope sling

Thimble

Eye hook with gate

Shackle

Two-leg bridle sling

Worksheet 3–6

Name _____   Date _____

Match the following fire extinguisher class with the symbol representing the type of fire.

Fire Type

Extinguisher Class

Class A

Class B

Class C

Class D

Worksheet 3–7

36 General Safety

Name _____ Date _____

Match the following U.S. Department of Transportation (DOT) hazardous material symbol with its classification and 49 CFR article number.

Hazard Material Symbol	**Classification/49 CFR Article**
	1/173.50
	2/173.115
	3/173.120
	4/173.124
	5/173.127
	6/173.132
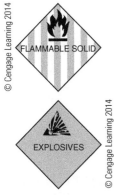	8/173.136

Worksheet 3–8

CHAPTER 4

Lubrication

Name _____ Date _____

Choose a term from the following to fill in the blank that completes the phrase describing an additive used to improve lubricant properties.

Corrosion inhibitor   Dispersants

Oxidation inhibitor   Antifoam

Demulsifiers   Viscosity index

Detergents   Extreme pressure

_____ are added to lubricants to break up deposits that form on metal components from exposure to elevated temperatures.

Pumping lubricant over gears and around bearing elements may allow air to become entrapped. _____ additives enable air bubbles to break up quickly, preventing damage from the disruption of the needed lubricant film.

_____ are added to lubricants to hold sludge and microscopic particles in suspension after a detergent additive breaks them down.

Contamination of lubricants with small amounts of water prevents the lubricant from forming a protective film. _____ enable water to separate from oil and allow lubricant to protect moving parts.

_____ are added to lubricants to improve their ability to stick to metals. The ability of a lubricant to provide a continuous protective film prevents air and moisture from contacting the metal and promoting oxidation of the surface.

_____ additives are used to improve the stability of a lubricant at elevated temperatures.

_____ additives react with a metal's surface to create a protective film. The purpose of the protective film is to improve load-carrying capacity and reduce the effects of shock loading.

_____ are added to lubricants to prevent breakdown of the hydrocarbon molecule during exposure to elevated temperatures in the presence of air.

Worksheet 4–1

Name _____   Date _____

Grease may be used in a variety of industrial applications. Selection of a grease formulation is typically made by the engineering group of an original equipment manufacturer (OEM) during the design phase of product development or by an experienced maintenance team. Use this exercise to develop an understanding of the grease-selection process. Choose a suitable grease formulation from the provided chart to match the bearing application described.

SKF bearing grease selection chart

Bearing working conditions	Temp	Speed	Load	Vertical shaft	Fast outer ring rotation	Oscillating movements	Severe vibrations	Shock load or frequent start-up	Low noise	Low friction	Rust inhibiting properties	Description	Temperature range (*1) LTL	Temperature range (*1) HTPL	Thickener / base oil	Base oil viscosity (*2)
LGMT 2	M	M	L to M	O	–	–	+	–	–	O	+	General purpose industrial and automotive	–30 °C / –22 °F	120 °C / 250 °F	Lithium soap / mineral oil	110
LGMT 3	M	M	L to M	+	O	–	+	–	–	O	O	General purpose industrial and automotive	–30 °C / –22 °F	120 °C / 250 °F	Lithium soap / mineral oil	120
LGEP 2	M	L to M	H	O	–	O	+	+	–	–	+	Extreme pressure	–20 °C / –4 °F	110 °C / 230 °F	Lithium soap / mineral oil	200
LGFP 2	M	M	L to M	O	–	–	–	–	–	O	+	Food compatible	–20 °C / –4 °F	110 °C / 230 °F	Aluminum complex / medical white oil	130
LGEM 2	M	VL	H to VH	O	–	+	+	+	–	–	+	High viscosity plus solid lubricants	–20 °C / –4 °F	120 °C / 250 °F	Lithium soap / mineral oil	500
LGEV 2	M	VL	H to VH	O	–	+	+	+	–	–	+	Extremely high viscosity with solid lubricants	–10 °C / 14 °F	120 °C / 250 °F	Lithium-calcium soap / mineral oil	1 020
LGLT 2	M to EH	L	L	O	–	–	–	O	–	+	O	Low temperature, extremely high speed	–50 °C / –58 °F	110 °C / 230 °F	Lithium soap / PAO oil	18
LGGB 2	L to M	L to M	M to H	O	–	+	+	+	–	O	O	Green biodegradable, low toxicity	–40 °C / –40 °F	90 °C (*3) / 194 °F	Lithium-calcium soap / synthetic ester oil	110
LGWM 1	L to M	L to M	H	–	O	+	–	+	–	–	+	Extreme pressure, low temperature	–30 °C / –22 °F	110 °C / 230 °F	Lithium soap / mineral oil	200
LGWM 2	L to M	L to H	M to H	O	O	O	+	+	–	–	+	High load, wide temperature	–40 °C / –40 °F	110 °C / 230 °F	Complex calcium sulphonate / synthetic (PAO)/mineral oil	80
LGWA 2	M to H	L to M	L to H	O	–	+	O	+	–	O	+	Wide temperature (*4), extreme pressure	–30 °C / –22 °F	140 °C / 284 °F	Lithium complex soap / mineral oil	185
LGHB 2	M to H	VL to M	H to VH	O	+	+	+	O	–	–	+	EP high viscosity, high temperature (*5)	–20 °C / –4 °F	150 °C / 302 °F	Complex calcium sulphonate / mineral oil	400
LGHP 2	M to H	M to H	L to M	+	–	–	O	O	+	O	+	High performance polyurea grease	–40 °C / –40 °F	150 °C / 302 °F	Di-urea / mineral oil	96
LGET 2	VH	L to M	H to VH	O	+	+	O	O	–	–	O	Extreme temperature	–40 °C / –40 °F	260 °C / 500 °F	PTFE / synthetic (fluorinated polyether)	400

+ = Recommended   O = Suitable   – = Not suitable

(*1) LTL = Low-temperature limit
HTPL = High-temperature performance limit
(*2) mm²/s at 40 °C / 104 °F = cSt

(*3) LGGB 2 can withstand peak temperatures of 120 °C / 250 °F
(*4) LGWA 2 can withstand peak temperatures of 220 °C / 428 °F

(*5) LGHB 2 can withstand peak temperatures of 200 °C / 392 °F

SKF Maintenance Products

Chapter 4 Lubrication

Name _____ Date _____

SKF bearing greases

- Highest quality grease for bearing lubrication
- Guarantee of consistent quality as each product is manufactured at one location to the same formulation
- A complete product programme for general and specific bearing lubrication requirements
- International standardisation of the SKF grease testing methods and equipment
- Worldwide product availability through the SKF dealer network

Available pack sizes

	SKF SYSTEM 24	35 g tube	200 g tube	420 ml cartridge	2 kg can	5 kg can	18 kg can	25 kg can	50 kg drum	180 kg drum	50 g (al-ml) syringe
LGMT 2		•	•	•	•	•	•		•	•	
LGMT 3		•	•	•	•	•	•		•	•	
LGEP 2			•	•	•	•	•		•	•	
LGFP 2		•	•		•	•	•				
LGEN 2			•		•	•	•				
LGEV 2			•		•	•					
LGLT 2		•		•	•		•				•
LGGB 2			•	•	•	•	•				
LGWM 1			•	•	•	•	•		•	•	
LGWM 2			•		•	•	•				
LGHB 2			•		•	•	•		•	•	
LGHP 2	•		•	•	•	•	•		•	•	
LGET 2			•		•		•				•

SKF Maintenance Products

Publication MP340EE - 2009/08
© SKF CARB, SYSTEM 24 are registered trademarks of the SKF Group

© SKF 2009

The contents of this publication are the copyright of the publisher and may not be reproduced (even extracts) unless prior written permission is granted. Every care has been taken to ensure the accuracy of the information contained in this publication but no liability can be accepted for any loss or damage whether direct, indirect or consequential arising out of the use of the information contained herein.

www.mapro.skf.com
skf.com/lubrication

Basic bearing grease selection

Generally use if: Speed = M, Temperature = M and Load = M	**LGMT 2** General purpose
Unless:	
Expected bearing temperature continuously > 100 °C / 212 °F	**LGHP 2** High temperature
Expected bearing temperature continuously > 150 °C / 302 °F, demands for radiation resistance	**LGET 2** Extremely high temperature
Low ambient -50 °C / -58 °F, expected bearing temperature < 50 °C / 122 °F	**LGLT 2** Low temperature
Shock loads, heavy loads, frequent start-up / shut-down	**LGEP 2** High load
Food processing industry	**LGFP 2** Food processing
"Green" biodegradable, demands for low toxicity	**LGGB 2** "Green" biodegradable

Note – For greases with relatively high ambient temperatures, use LGMT 3 instead of LGMT 2
– For special operating conditions, refer to the SKF bearing grease selection chart

Bearing operating parameters

Temperature
- L = Low
- M = Medium
- H = High
- EH = Extremely high

Speed for ball bearings
- EH = Extremely High
- VH = Very High
- H = High
- M = Medium
- L = Low

Speed for roller bearings
- H = High
- M = Medium
- L = Low
- VL = Very Low

Load
- VH = Very high
- H = High
- M = Medium
- L = Low

Temperature	
<50 °C / 122 °F	
50 to 100 °C / 122 to 230 °F	
>100 °C / 212 °F	
>150 °C / 302 °F	

Speed (ball)	n.dm over 700 000
	n.dm up to 700 000
	n.dm up to 500 000
	n.dm up to 300 000
	n.dm below 100 000

Speed (SRB/TRB/CARB)	CRB
n.dm over 210 000	n.dm over 270 000
n.dm up to 210 000	n.dm up to 270 000
n.dm up to 75 000	n.dm up to 75 000
n.dm below 30 000	n.dm below 30 000

Load	
C/P < 2	
C/P < 4	
C/P < 8	
C/P < 15	

SKF Maintenance Products

Worksheet 4–3

40 General Knowledge

Name _____ Date _____

Possible grease formulations:

LGMT2, LGEP2, LGLT2, LGWM1, LGWM2, LGHB2, LGHP2, and LGET2

Bearing Application Summary

Application	Temperature	Bearing Speed	Load	Severe Vibration	Vertical Shaft	Corrosion Issue	Shock Load
1	M	M	M	Yes	Yes	Yes	No
2	L	L	H	No	No	Yes	Yes
3	H	H	L	No	Yes	Yes	Yes
4	VH	L	VH	Yes	No	No	No
5	M	M	H	Yes	Yes	Yes	Yes
6	M	L	H	Yes	No	Yes	Yes
7	L	EH	L	No	No	No	No
8	H	VL	VH	Yes	Yes	Yes	Yes

Grease application 1 _____LGMT 2_____

Grease application 2 _____

Grease application 3 _____

Grease application 4 _____

Grease application 5 _____

Grease application 6 _____

Grease application 7 _____

Grease application 8 _____

Name _____   Date _____

Many bearing assemblies may be constructed with ball or roller moving elements to reduce friction between the rotating shaft and the stationary component supporting the shaft. Other bearings may be made of solid machined or sintered metal cylindrical components to support the rotating shaft. This bearing type is considered a plain bearing or a journal bearing. Journal bearings count on a thin layer of lubricant to separate the moving shaft from the stationary bearing. Discussions in the textbook indicate a full film lubrication condition is ideal to minimize wear of the components and reduce system losses from friction. This exercise will introduce a reference chart to assist in determining the SAE lubricant viscosity grade given the shaft speed (RPM) and the relative load supported by the shaft and bearing. Use the provided chart to answer questions on the following page.

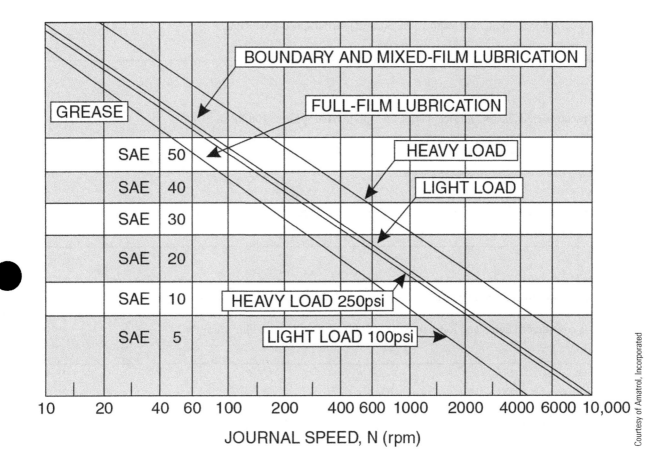

Worksheet 4–5

42 General Knowledge

Name _____ Date _____

Determine the lubrication condition for the following scenarios. Indicate the condition for each scenario as *boundary and mixed film* or *full film*. Follow the lines for each of the variables on the chart to determine where they intersect. Use this intersection location to determine the lubricant condition.

1. Shaft load is considered "light," lubricant is SAE 30, and journal speed is 400 RPM.

   This scenario indicates boundary and mixed film condition.

2. Shaft load is considered "light," lubricant is SAE 50, and journal speed is 60 RPM.

   _____

3. Shaft load is considered "heavy," lubricant is grease, and journal speed is 100 RPM.

   _____

4. Shaft load is considered "heavy," lubricant is SAE 5, and journal speed is 2000 RPM.

   _____

Use the same chart now to determine the correct lubricant grade for each of the following scenarios.

1. Shaft load is considered "light," and journal speed is 3,000 RPM with a full film condition.

   SAE 5.

2. Shaft load is considered "heavy," and journal speed is 600 RPM with a full film condition.

   _____

3. Shaft load is considered "light," and journal speed is 200 RPM with a mixed film condition.

   _____

4. Shaft load is considered "heavy," and journal speed is 1,500 RPM with a full film condition.

   _____

Worksheet 4–6

Name _____   Date _____

A comprehensive preventative maintenance (PM) program is important to ensure expensive equipment is fully operational whenever it is needed for production. The addition of a predictive maintenance program goes beyond inspections and replenishing lubricants. Predictive maintenance uses information collected during the preventative maintenance cycle to predict when equipment should be overhauled before an unscheduled shutdown or catastrophic failure occurs. Review the following oil-analysis data and determine at which PM cycle the hydraulic equipment should be removed from service or scheduled for an overhaul. Use Appendix E, Maintenance Fundamentals for Wind Technicians, as a guide to recommended minimum and maximum values.

PM Date	11/10/09	5/05/10	11/22/10	5/11/11	10/30/11	5/02/12	11/08/12
Viscosity	42.0	42.1	43.0	43.5	44.2	44.3	44.5
TAN	0.54	0.58	0.56	0.58	0.57	0.59	0.64
Metals (ppm)							
Iron	2	5	8	16	18	22	38
Aluminum	0	0	0	1	0	0	0
Chromium	0	1	1	0	2	2	8
Copper	5	4	5	6	5	8	16
Lead	0	0	0	1	1	2	6
Tin	0	0	0	0	0	1	4
Nickel	0	1	1	1	2	2	3
Silver	0	2	2	3	3	3	4
Silicon	4	5	8	9	12	13	14
Other Contaminants (ppm)							
Water	50	53	53	122	126	131	135

During which PM cycle should the hydraulic system shown here be removed from service for a thorough review?
_____
_____

Why? _____
_____
_____
_____
_____

Worksheet 4–7

CHAPTER 5

Fluid Power

Name _____   Date _____

1. Calculate the pressure change required to compress 10 ft³ of air into a 1-ft³ pneumatic system reservoir. Use original air pressure (P_1) of 15 psi. Disregard changes in temperature resulting from compression processes for each calculation.

2. Calculate the original air volume (V_1) for a compressor system with initial air intake pressure of 15 psi (P_1) and final reservoir conditions of 100 psi (P_2) and 5,184 in.³ (V_2).

3. Calculate the volume change (ΔV) for a pneumatic system with a pressure increase of 345 kPa (ΔP). Final answer should be in units of m². *Remember:* 1 Pa = 1 N/m².

4. Calculate the percent volume change in a hydraulic fluid when exposed to an increase in system pressure of 60,000 kPa. Use a bulk modulus value of 250,000 psi for the system hydraulic fluid.

Worksheet 5–1

46   General Knowledge

Name _____   Date _____

Use the following fluid system diagram to determine parameters listed for each question.

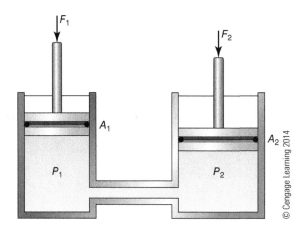

1. Determine area (A_2) of piston 2 that will produce a force (F_2) of 1,000 lb$_f$ with the following system values:

   A_1 = 3 in.²

   F_1 = 500 lb$_f$

   F_2 = 1,000 lb$_f$

   A_2 = _____

2. Determine force (F_1) applied on piston 1 that will produce a fluid pressure in cylinder 2 of 1,500 psi with the following system values:

   A_1 = 0.5 in.²

   A_2 = 1.5 in.²

   F_1 = _____

Worksheet 5–2

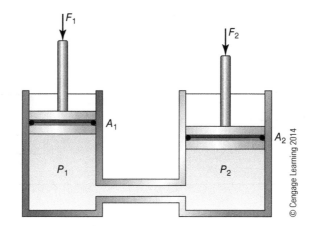

1. Determine area (A_2) for piston 2 that will produce a force (F_2) of 1,000 N with the following system values:

   A_1 = 650 mm²

   F_1 = 7,250 N

   F_2 = 10,000 N

   A_2 = _____

2. Determine the force (F_1) applied on piston 1 that will produce a pressure in cylinder 2 of 70 bar with the following values:

   A_1 = 5 in.²

   A_2 = 7.5 in.²

   P_2 = _____

Worksheet 5–3

48 General Knowledge

Name _____ Date _____

Match a hydraulic system function in the following with each device shown as an ANSI symbol.

Energize fluid

Control

Convert fluid power to output

Support

_____ _____

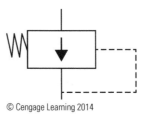

_____ _____

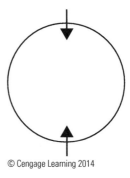

_____ _____

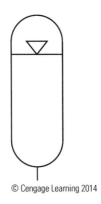

_____ _____

Worksheet 5–4

Name _____   Date _____

1. Complete the following hydraulic circuit with a pressure-relief valve.

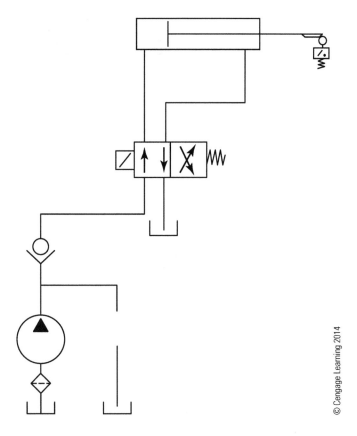

Complete the following hydraulic circuit with a three-way, two-position solenoid valve.

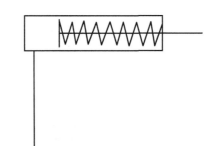

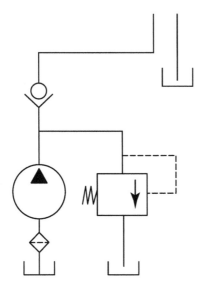

Name _____ Date _____

Complete the following hydraulic circuit with a variable pressure-reducing valve.

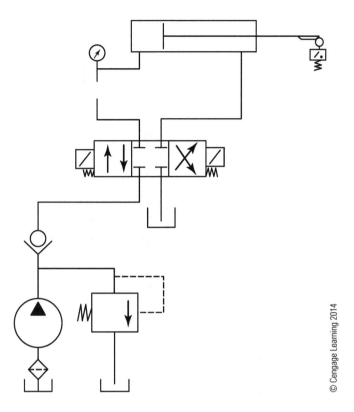

Worksheet 5-7

General Knowledge

Name _____ Date _____

Complete the following hydraulic circuit with a bidirectional hydraulic motor.

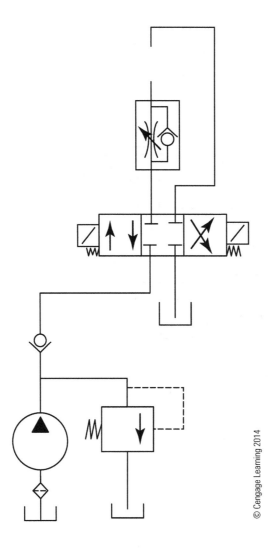

Name _____   Date _____

Complete the hydraulic circuit below with a four-way, three-position solenoid valve (left position = flow extends rod, center position = ports blocked, and right position = flow retracts rod) and an accumulator for emergency closure of the brake assembly.

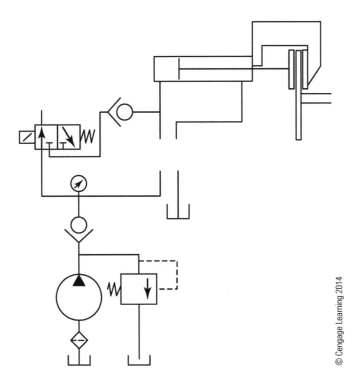

Worksheet 5–9

General Knowledge

Name _____   Date _____

Calculate the torque acting on the bolt for the following hydraulic power wrench parameters:

Piston area (A_1) = 1 in.²

System pressure (P_1) = 6,000 psi

Torque arm (L) = 2 in.

Torque output _____

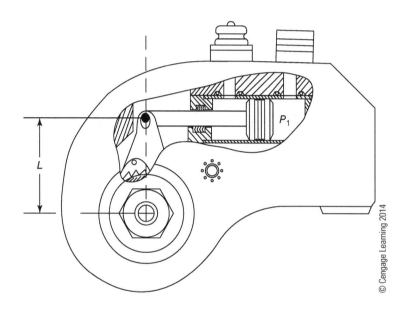

Worksheet 5–10

CHAPTER 6

Bolting Practices

Name _____ Date _____

Identify the bolt features on the following image from the list of possible terms.

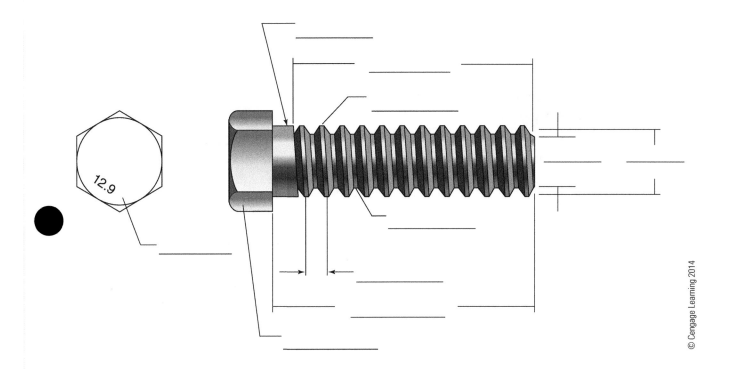

Grade or class marking
Shoulder
Crest
Minor diameter
Major diameter

Head
Root
Thread length
Fastener length
Pitch

Worksheet 6–1

General Knowledge

Name _____ Date _____

Identify the following fastener component from the list of possible terms.

Lock nut	Flat washer
Wing nut	Hex nut
Shake-proof washer	Square nut
Wing nut	E clip
Lock or split washer	Roll pin
Retaining ring (external and internal)	Cotter pin
Socket-head cap screw	Hex-head machine screw
Sheet-metal screw	Stud

_____

_____

_____

_____

_____

_____

_____

_____

Worksheet 6–2

Name _____ Date _____

Chapter 6 Bolting Practices 57

_____

_____

_____

_____

_____

_____

_____

Worksheet 6–3

58 General Knowledge

Name _____ Date _____

Identify the fastener class or grade marking along with organization that sets the standard specification for the marking of the following.

_____ ASTM Grade A325 Type 3 _____

_____

_____

_____

_____

_____

_____

_____

_____

Worksheet 6–4

Name _____   Date _____

1. Identify the regions and points on the stress–strain curve using the following terms.
   A. Proportional limit
   B. Elastic limit
   C. Yield strength
   D. Ultimate strength
   E. Modulus of elasticity
   F. Rupture strength
   G. Elastic deformation
   H. Plastic deformation

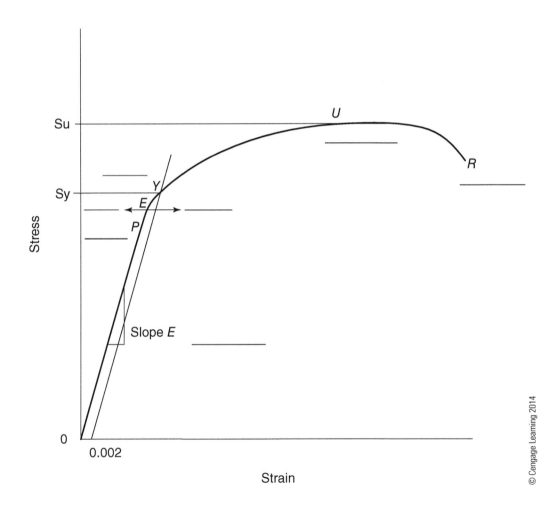

Worksheet 6–5

Name _____   Date _____

Calculate the elongation (δ) of a hex bolt subjected to a 170,000-N tensile load given the following fastener–joint assembly parameters.

Fastener cross-sectional area (A) = 910 mm^2

Modulus of elasticity (E) = 207,000 MPa

Length of fastener between head and nut (L) = 152 mm

Thread pitch = 3 mm

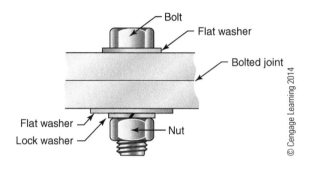

Name _____   Date _____

Calculate the tensile load (F) on the fastener for each angle of rotation of the nut given the following fastener–joint assembly parameters.

Load (F) at nut angle 0° = 26,000 N

Fastener cross-sectional area (A) = 707 mm²

Modulus of elasticity (E) = 207,000 MPa

Length of fastener between head and nut (L) = 200 mm

Thread pitch = 2 mm

Position 1, 45° rotation; F = _____

Position 2, 90° rotation; F = _____

Position 3, 135° rotation; F = _____

Position 4, 225° rotation; F = _____

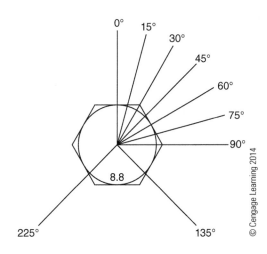

Worksheet 6–7

62 General Knowledge

Name _____   Date _____

Calculate the fastener torque level to be supplied by a power bolting tool for the following assembly conditions.

1. Bolt size: 1½–6 UNC

   Material: SAE Grade 5

   Surface finish: Plain steel, waxed

   Clamp force: 78,000 lb_f

   Torque (T) = _____ (ft-lbs)

2. Bolt size: 2–4½ UNC

   Material: SAE Grade 8

   Surface finish: Plain steel, slightly oiled

   Clamp force: 225,000 lb_f

   Torque (T) = _____ (ft-lbs)

3. Bolt size: 3–4 UNC

   Material: ASTM A193 B7

   Surface finish: Hot dip, galvanized

   Clamp force: 425,363 lb_f

   Torque (T) = _____ (ft-lbs)

4. Bolt size: M20–2.5

   Material: 8.8

   Surface finish: Hot dip, galvanized

   Clamp force: 96,000 N

   Torque (T) = _____ (N-m)

Worksheet 6–8

CHAPTER 7

Test Equipment

Name _____   Date _____

Match each of the following standard electrical symbols with its corresponding name.

Contactor coil

Power supply

Heater

Wires connected

Time-delay relay coil

Ground potential

Motor

Terminal strip

Circuit breaker

Worksheet 7–1

64 General Knowledge

Name _____ Date _____

Match each of the following standard electrical symbols with its corresponding name.

Battery

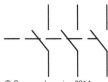

Pushbutton NO

Fuse

Wires not connected

Shielded conductors

Pushbutton NC

Thermal overload device

Switch

Multiple conductors in cable

Worksheet 7–2

Name _____   Date _____

Write the corresponding name below each of the following standard electrical symbols.

Potential transformer	Transformer (Δ-Y connections)	Transformer
Zener diode	Inductor (air core)	Resistor (fixed)
Inductor (multiple taps)	Transformer (iron core)	Current transformer

_____

_____

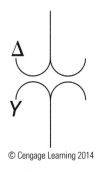

_____

_____

_____

_____

_____

Worksheet 7–3

Name _____   Date _____

Write the corresponding name below each of the following standard electrical symbols.

Liquid level switch (NC)	Pressure switch (NC)	Limit switch (NO)
Relay contact pair (NO–NC)	Proximity switch (NC)	Temperature switch (NO)
Light Emitting Diode (LED)	Rectifier (full bridge)	Flow switch (NO)

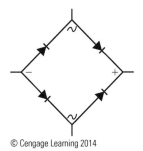

_____

_____

_____

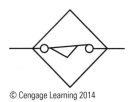

_____

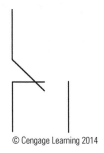

_____

Worksheet 7–4

Name _____   Date _____

Write the corresponding test equipment name below each of the following instruments.

Power-quality meter	Infrared thermometer	Two-channel digital O-scope
Two-channel O-scope	Current meter (amp clamp)	Bore scope
Multi meter	Infrared camera	Vibration meter

_____

_____

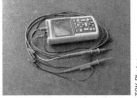

_____

_____

_____

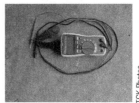

_____

_____

_____

_____

Worksheet 7–5

68 General Knowledge

Name _____   Date _____

Determine the peak, peak to peak, and frequency for the following oscilloscope setup and signal description.

1. Setup: Channel voltage setting is 2 mV/division, time setting is 10 mS/division, and probe is set on 1X.

    Signal description: Signal extends 6.2 divisions from peak to peak, and the distance for one complete cycle is 4.6 divisions.

    Peak _____

    Peak to peak _____

    Frequency _____

2. Setup: Channel voltage setting is 5V/division, time setting is 5 mS/division, and probe is set on 1X.

    Signal description: Signal extends 3.5 divisions from peak to peak, and the distance for one complete cycle is 5 divisions.

    Peak _____

    Peak to peak _____

    Frequency _____

3. Setup: Channel voltage setting is 2 mV/division, time setting is 10 mS/division, and probe is set on 1X.

    Signal description: Signal extends 5.1 divisions from peak to peak, and the distance for one complete cycle is 2.5 divisions.

    Peak _____

    Peak to peak _____

    Frequency _____

Worksheet 7–6

Chapter 7   Test Equipment

Name _____   Date _____

Determine the electrical component located at 4-C of the following electrical schematic.

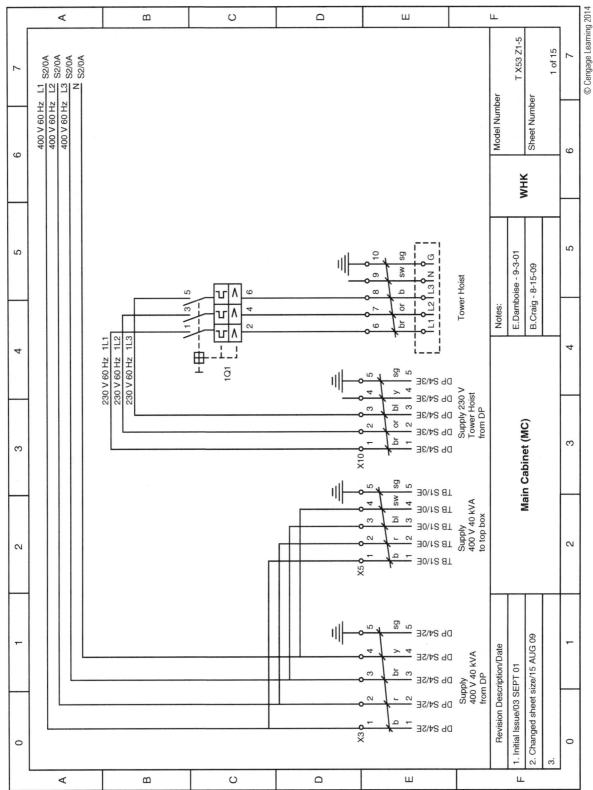

Worksheet 7–7

General Knowledge

Name _____   Date _____

The terminal strip located at 0-E of the following schematic has wires attached from a down tower control cabinet. What is the schematic sheet and location listed for the supply connections of the down tower cabinet terminal strip? Worksheet 7-7 shows an example of a down tower control cabinet schematic sheet.

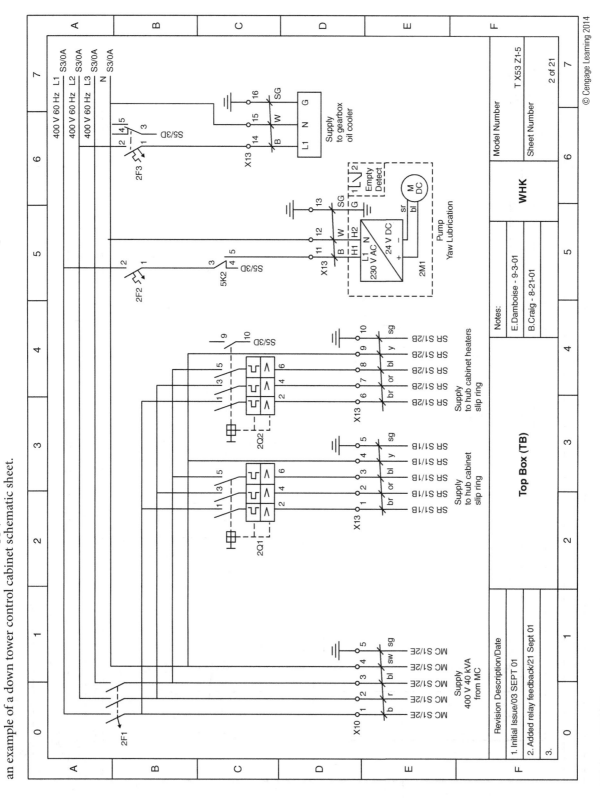

Worksheet 7-8   Page 1 of 2

Name _____   Date _____

What is the voltage indicated for the top conductor shown at 5-C of the electrical following schematic?

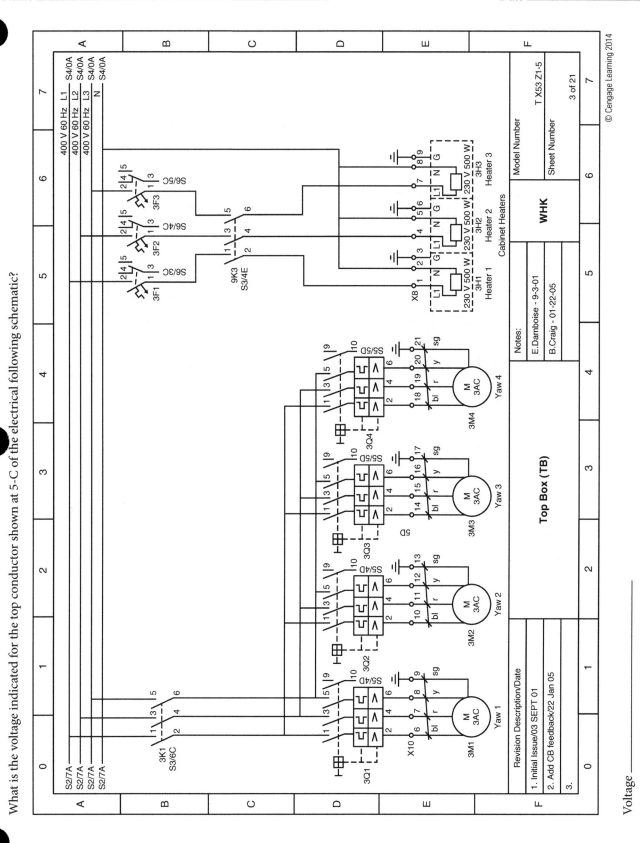

Voltage _____

Worksheet 7-9

General Knowledge

Name _____   Date _____

Determine the voltage potential between conductors shown at 6-D and the device located at 4-E of the following electrical schematic.

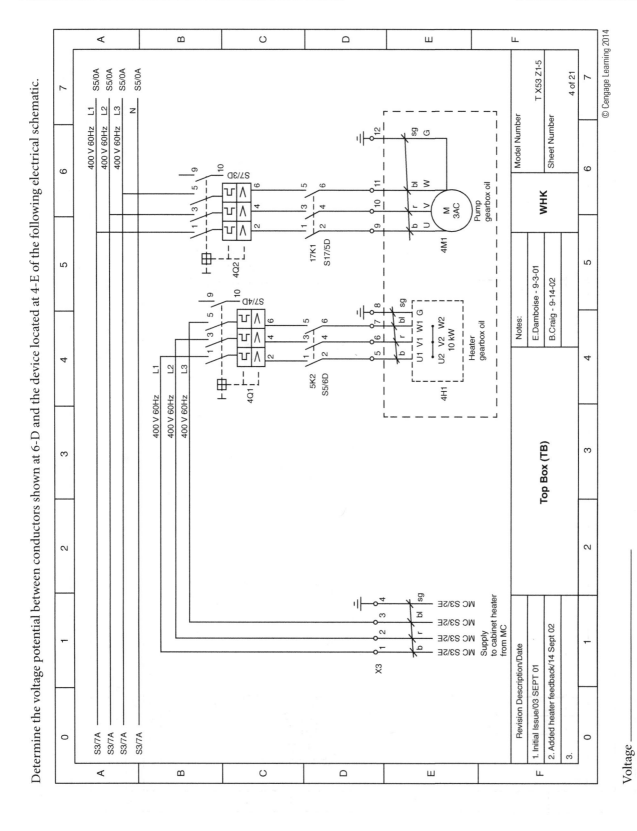

Voltage _____

Device _____

Worksheet 7–10

CHAPTER 8

Component Alignment

Name _____ Date _____

Write the correct term by the drive-train component in the following diagram.

A. Rotor assembly
B. Rotor adapter
C. Main bearing
D. Main shaft
E. Compression coupling
F. Gearbox
G. Brake assembly
H. High-speed coupling

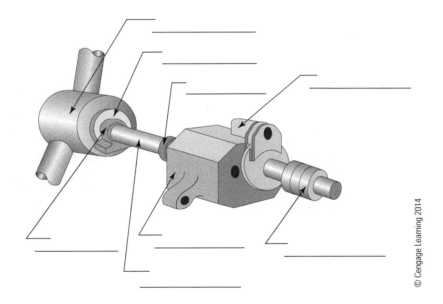

Worksheet 8-1

74 General Knowledge

Name _____ Date _____

Write the gear-type name listed in the following with the corresponding gear image.

Bevel gear

Spur gear

Worm gear

Helical gear

Herringbone gear

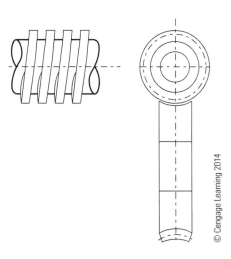

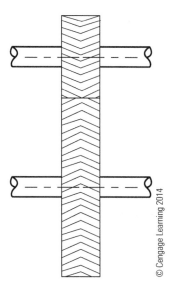

_____ _____

_____ _____

_____

Worksheet 8–2

Name _____   Date _____

● Calculate the gear ratio for each of the following gear sets.

Gear Ratio: _____

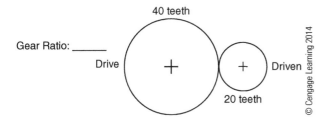

_____

Gear Ratio: _____

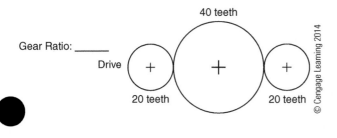

_____

Gear Ratio: _____

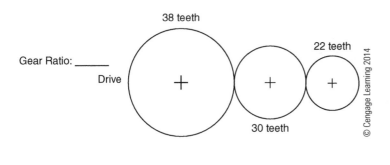

_____

●

Worksheet 8–3

76 General Knowledge

Name _____ Date _____

Gear Ratio: _____

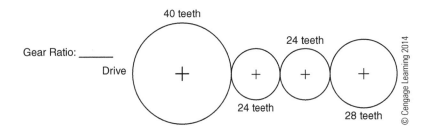

Gear Ratio: _____

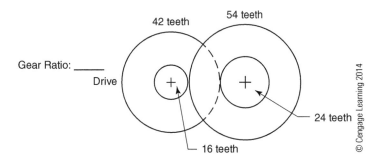

Gear Ratio: _____

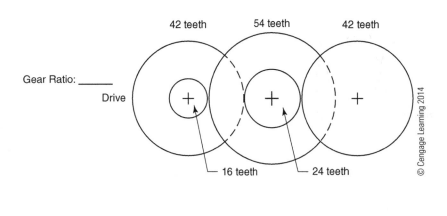

Worksheet 8–4

Name _____ Date _____

Determine the output gear direction of rotation, velocity (RPM), and torque given the gear drive-train assembly parameters, input velocity, and torque.

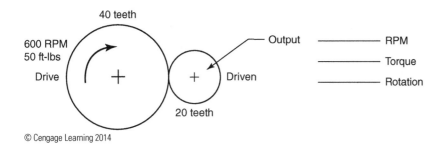

_____ RPM
_____ Torque
_____ Rotation

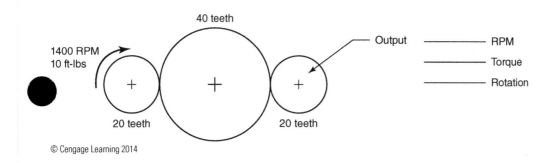

_____ RPM
_____ Torque
_____ Rotation

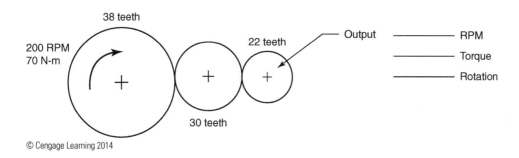

_____ RPM
_____ Torque
_____ Rotation

Worksheet 8–5

General Knowledge

Name _____ Date _____

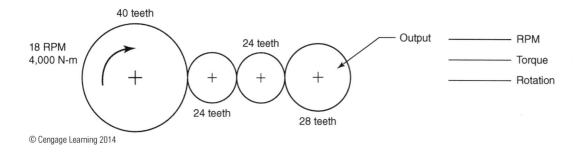

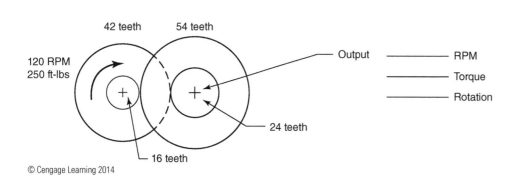

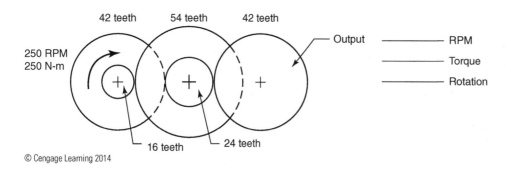

Worksheet 8–6

Name _____   Date _____

- Indicate the type of shaft misalignment shown by the following diagrams.

Axial

Parallel or radial

Angular or offset

Correct alignment

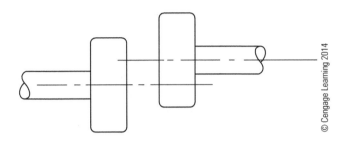

_____

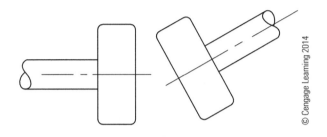

_____

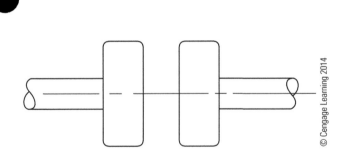

_____

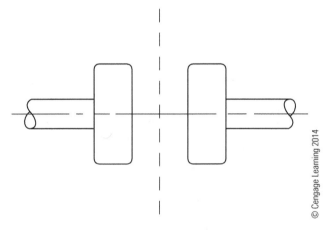

_____

Worksheet 8–7

80 General Knowledge

Name _____   Date _____

Indicate the type of soft foot condition shown with the following diagrams.

Induced

Angular

Parallel

Springing

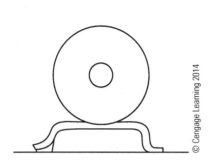

_____

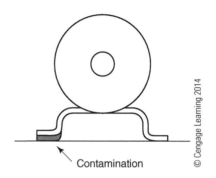

Contamination

_____

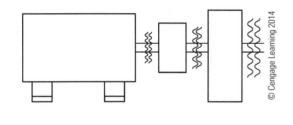

_____

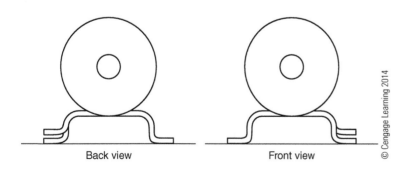

Back view   Front view

_____

Worksheet 8–8

Name _____   Date _____

Determine the minimum number of each shim size required to fill a gap between the generator foot and the generator mounting assembly to correct for a misalignment or soft foot condition.

List the number of each shim required beside the following standard sizes.

1. Gap: 0.78 mm   0.001" __1__   0.005" __0__   0.010" __1__

   0.020" __1__   0.050" __0__   0.100" __0__

2. Gap: 0.473"   0.001" __3__   0.005" __0__   0.010" __0__

   0.020" __1__   0.050" __1__   0.100" __4__

3. Gap: 0.51 mm   0.001" __0__   0.005" __0__   0.010" __0__

   0.020" __1__   0.050" __0__   0.100" __0__

4. Gap: 0.224"   0.001" __4__   0.005" __0__   0.010" __0__

   0.020" __1__   0.050" __0__   0.100" __2__

5. Gap: 1.75 mm   0.001" __4__   0.005" __1__   0.010" __1__

   0.020" __0__   0.050" __1__   0.100" __0__

82 General Knowledge

Name _____ Date _____

Determine the amount of horizontal or vertical movement required to correct a misaligned generator given the information provided for each scenario.

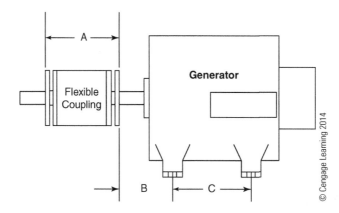

1. Horizontal–angular misalignment: Determine the horizontal movement of the rear feet if the front feet are indicated to be offset to the left by 2.5 cm on the laser alignment-control unit. *Hint:* Movement of the rear feet can be calculated using similar right triangles with the base length associated with the front feet being 2.5 cm and the height approximated as dimension B to determine the offset angle. Using this offset angle and the new height as B + C, determine the movement as the larger right-triangle base dimension.

    Direction _____ Distance _____

2. Horizontal–parallel misalignment: Determine the horizontal movement of the front feet if the rear feet are indicated to be offset to the right by 5 cm on the laser alignment-control unit.

    Direction _____ Distance _____

3. Vertical–angular misalignment: Determine the vertical movement of the front feet if the rear feet are indicated to be offset downward by 6.5 cm on the laser alignment-control unit.

    Direction _____ Distance _____

4. Vertical–parallel misalignment: Determine the horizontal movement of the front feet if the rear feet are indicated to be offset to the left by 2 cm on the laser alignment-control unit.

    Direction _____ & Distance _____

Worksheet 8–10

CHAPTER

9 Down Tower Assembly

Name _____   Date _____

Use the electrical schematics for Main Cabinet (sheet 1) and Distribution Panel (sheet 4) to determine the circuit breaker or fused disconnect switch used to isolate and LOTO Main Cabinet terminal block X3 to verify the torque value of fasteners connecting power lines L1–L2–L3.

Circuit breaker (CB) or fused disconnect switch (F): _____

Phase-to-phase voltage on these lines: _____

Worksheet 9–1

84 Preventative Maintenance

Name _____ Date _____

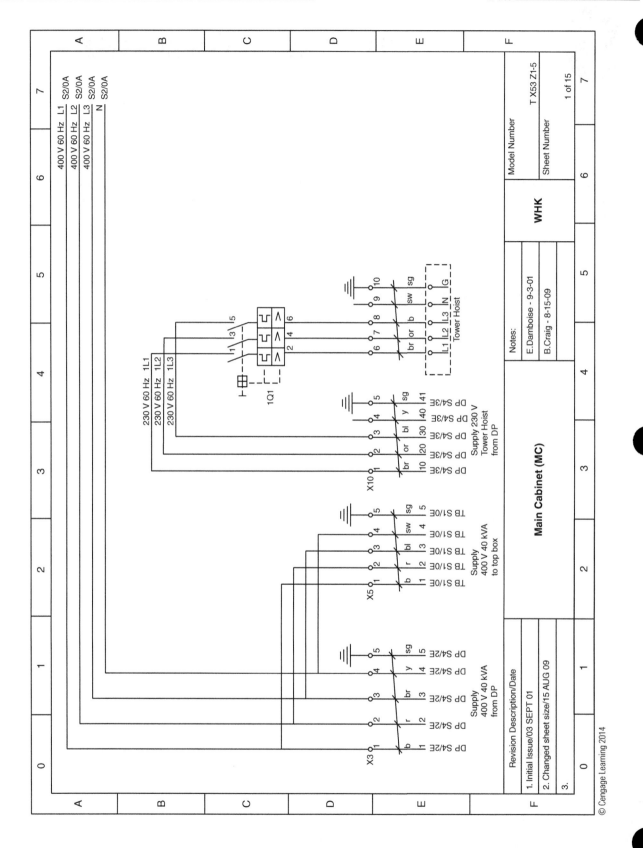

Name _____   Date _____

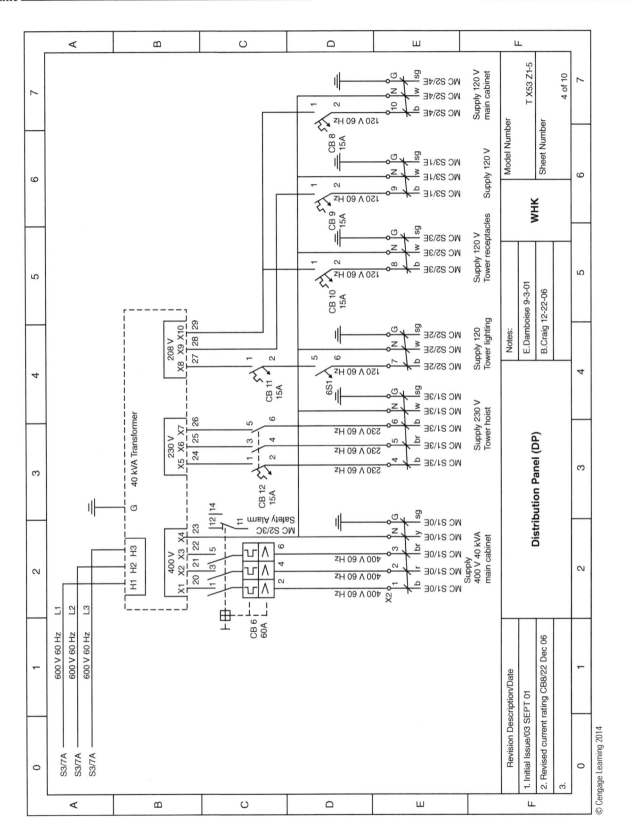

Worksheet 9-3

86 Preventative Maintenance

Name _____   Date _____

The power-converter system for the wind turbine you have been assigned to maintain uses a 50:50 ethylene glycol–water solution. Look up and review an MSDS for this chemical and determine the following requirements.

Manufacturer's product name: _____

List the two major components of the solution and associated CAS numbers.

1. Name _____ CAS Number _____

2. Name _____ CAS Number _____

Action to be taken for accidental ingestion: _____

_____

_____

Recommended PPE: _____

Boiling point: _____

Freezing point (50:50 solution): _____

UN ID number (label for bulk or large shipments): _____

DOT hazard class (label for bulk or large shipments): _____

Contact: _____

Phone number: _____

Worksheet 9–4

Name _____   Date _____

It is important to ensure that the cooling solution in a liquid-cooled power-conversion system is maintained to the proper solution ratio to prevent freezing during cold days when the system is off-line and boiling over when under load. This exercise shows the relationship between solution mixture and working temperatures.

Materials and Supplies

1 quart of premixed 50:50 ethylene glycol and water

1 gallon of distilled water

Clean 1 gallon container for mixing

Hydrometer test instrument showing coolant working temperature values (see Figure 9-6 in the MFWT textbook).

Procedure

1. Review the MSDS sheet for the coolant solution brand to be used for the exercise. Note any safety items for first aid and use the recommended PPE.

   _____

   _____

2. Pour 1 pint of premixed 50:50 solution into the empty mixing container. Draw fresh solution into the hydrometer test instrument. Determine the relative freezing and boiling point of the fresh coolant solution from the test instrument indicator.

   Freezing temperature: _____

   Boiling temperature: _____

3. Dilute the 50:50 coolant solution with distilled water. Pour 1 pint of distilled water into the 50:50 premixed solution. This will produce a 25:75 solution (1 part ethylene glycol and 3 parts water). Determine the relative freezing and boiling temperatures.

   Freezing temperature: _____

   Boiling temperature: _____

4. Dilute the 25:75 coolant solution with distilled water. Pour 1 pint of distilled water into the 25:75 solution. This will produce a 16.7/83.3 solution (1 part ethylene glycol and 5 parts water). Determine the relative freezing and boiling temperatures.

   Freezing temperature: _____

   Boiling temperature: _____

88 Preventative Maintenance

Name _____   Date _____

5. Use the following graph to plot the freezing and boiling temperatures with respect to the solution mixture ratio. Temperature is listed on the left side axis of the graph, and mixture ratio is listed along the bottom axis.

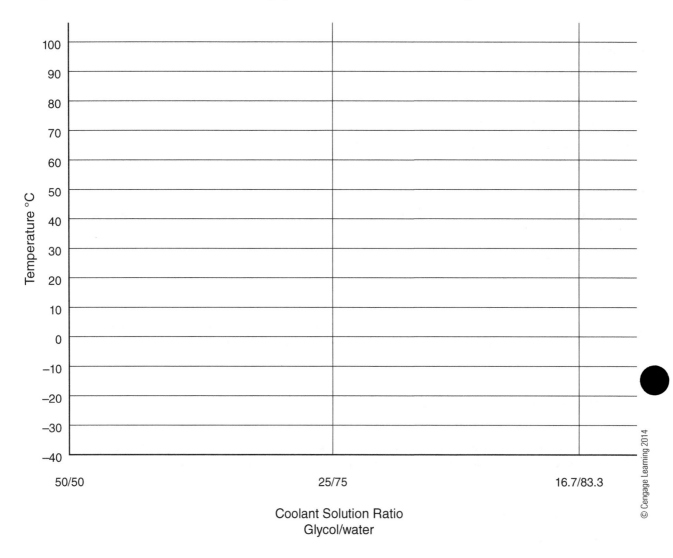

6. How did the freezing and boiling temperatures change with the addition of distilled water?

_____

_____

The addition of distilled water to a coolant system instead of the prescribed premixed 50:50 solution will have the same effect as seen here. It is very important to follow an approved procedure or manufacturer's recommendations when adding coolant to the reservoir.

Worksheet 9–6

Name _____ Date _____

Match the torque-inspection tool name from the following list with the correct diagram of the tool.

Torque Tool Type **Torque Tool Accessory**

Beam Multiplier

Dial Angle gauge

Micrometer Adapter

Screwdriver

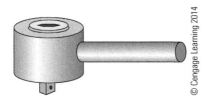
_____

_____

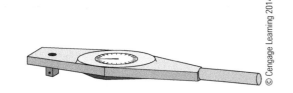
_____

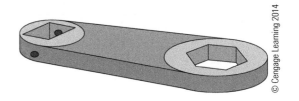
_____

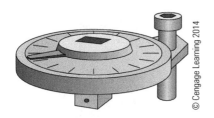
_____

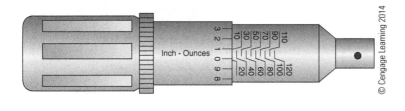
_____

_____

Worksheet 9–7

90 Preventative Maintenance

Name _____ Date _____

Determine the torque value shown on each of the following micrometer torque-wrench barrel diagrams.

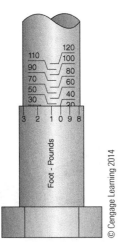

_____

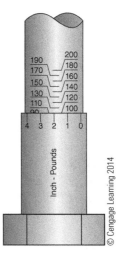

_____

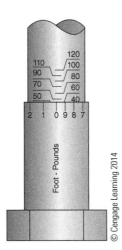

_____

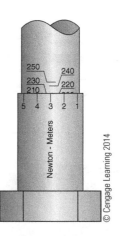

_____

Worksheet 9–8

Name _____   Date _____

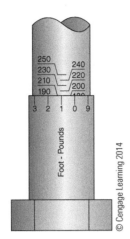

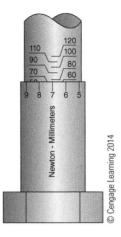

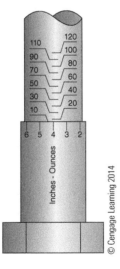

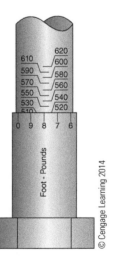

Worksheet 9–9

CHAPTER 10

Tower

Name _____ Date _____

Use electrical schematics on the next two pages to determine the circuit breaker(s) needed to isolate and LOTO tower-lighting, tower-receptacle, and tower-hoist circuits.

Tower-lighting circuit breaker (CB): _____

Schematic page location: _____

Tower-receptacle circuit breaker (CB): _____

Schematic page location: _____

Voltage level of the circuit: _____

Tower-hoist circuit breaker (CB): _____

Schematic page location: _____

Voltage level of the circuit: _____

Worksheet 10–1

94 Preventative Maintenance

Name _____   Date _____

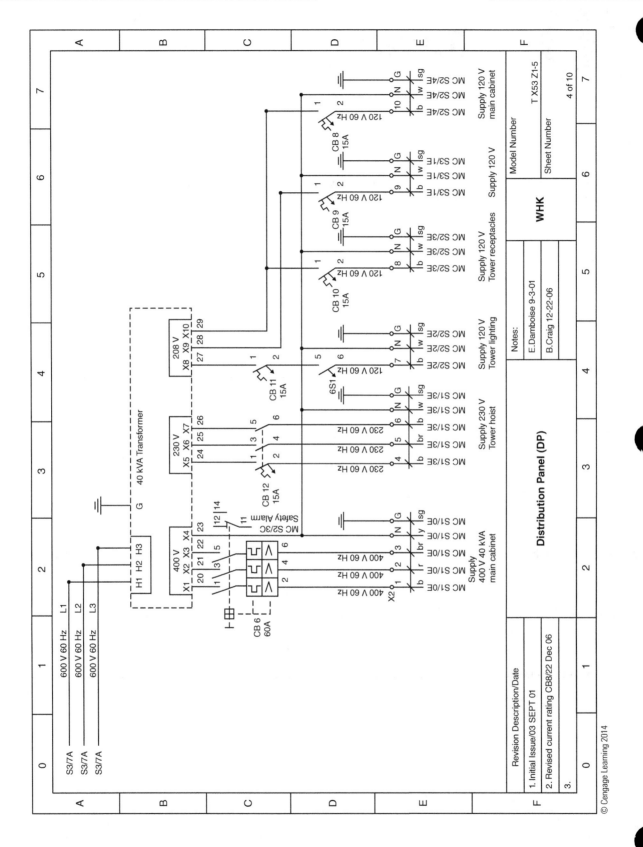

Worksheet 10–2

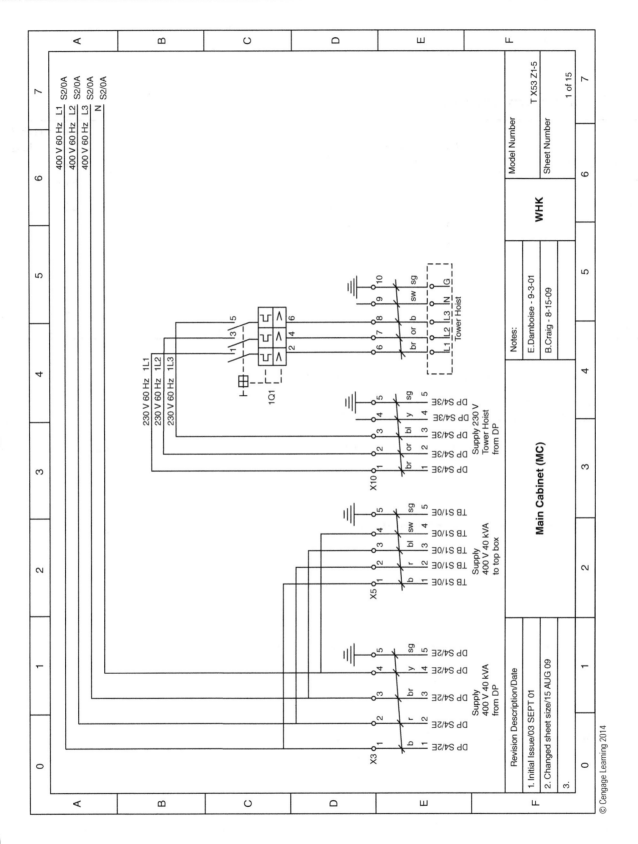

Preventative Maintenance

Name _____   Date _____

Determine the requested information for wind-turbine power cables given the conductor property estimates and information listed for each of the following problems.

Conductor Property Estimates

Conductor Size	DC Resistance @ 75°C (167°F) Coated Copper		Insulated Cable Approximate Weight	Convert lbs/ft to kg/m in the Column Below
kcmil	Ohm/km	Ohm/kft	lbs/ft	kg/m
500	0.0869	0.0265	1.74	2.589
600	0.0732	0.0223	2.09	_____
700	0.0622	0.0189	2.44	_____
750	0.0579	0.0176	2.61	_____
800	0.0544	0.0166	2.78	_____
900	0.0481	0.0147	3.13	_____
1000	0.0434	0.0132	3.48	_____

Formula for resistance change because of change in conductor temperature:

$R_2 = R_1[1 + \alpha(T_2 - 75)]$;
$\alpha_{cu} = 0.00323$ is the temperature coefficient for resistance of copper.

Information listed in this table is for reference only. Consult electrical standards such as NFPA 70 or engineering handbooks for exact values.

1. Determine the mass (kg) of 80 meters of 500-kcmil insulated copper cable. _____ 207.12 kg _____

2. Determine the mass (kg) of 100 meters of 1,000-kcmil insulated copper cable. _____

3. Determine the DC resistance (@55°C) for 300 feet of 700-kcmil coated copper conductor.

   _____

4. Determine the DC resistance (@95°C) for 300 feet of 700-kcmil coated copper conductor.

   _____

5. Determine the DC resistance (@75°C) for three 300-ft sections of 700-kcmil coated copper conductor connected in parallel. _____

Worksheet 10-4

Chapter 10 Tower

Name _____   Date _____

Determine the minimum tensile and yield strength for the following ISO class fasteners.

1. Bolt: M22 × 2.5

   Class: 10.9

   Minimum tensile strength: _____1,000_____ MPa
   Minimum yield strength: _____900_____ MPa

   Example: Class 10.9 fastener—minimum tensile strength 10 × 100 = 1,000 MPa and 0.9 (90%) is the relationship between minimum yield strength and the minimum tensile strength, so minimum yield strength for this fastener would be 1,000 MPa × 0.9 = 900 MPa.

2. Threaded stud: M30 × 3.5

   Class: 12.9

   Minimum tensile strength: _____ MPa
   Minimum yield strength: _____ MPa

3. Hex head machine screw: M16 × 2.0

   Class: 8.8

   Minimum tensile strength: _____ MPa
   Minimum yield strength: _____ MPa

4. Threaded stud M10 × 1.5

   Class: 5.8

   Minimum tensile strength: _____ MPa
   Minimum yield strength: _____ MPa

5. Threaded stud M8 × 1.25

   Class: 4.6

   Minimum tensile strength: _____ MPa
   Minimum yield strength: _____ MPa

Worksheet 10–5

98 Preventative Maintenance

Name _____   Date _____

Determine the torque [N-m] to be applied for the following fastener assembly scenarios.

1. Fastener: M22 × 2.5

   Class: 10.9

   Stress area: 303.4 mm²

   Surface finish: Hot dip, galvanized

   Permanent assembly loaded to 90% of fastener yield strength.

   Torque: _____1,352_____ [N-m]

Stress (σ) = F/A, so clamp force (F) = $\sigma \times$ A = 900 [MPa] × 303.4 [mm²]; 1 MPa = 1 N/mm²

Clamp force (F) = $\sigma \times$ A = 900 [N/mm²] × 303.4 [mm²] = 273,060 N

To ensure the stress created in the bolt never exceeds 90% of the yield strength, use 90% of the calculated clamp force: clamp force (F) = 273,060 N (0.90) = 245,754 N.

Equation for torque: [N-m] (T) = KDF/1000; K is coefficient for surface finish, D is nominal diameter of fastener, and F is the clamp force applied to the joint by the fastener. There are 1,000 millimeters in a meter.

T [N-m] = (0.25 × 22 mm × 245,754 N)/(1,000 mm/m) = 1,351.65 N-m

2. Fastener: M30 × 3.5

   Class: 10.9

   Stress area: 560.5 mm²

   Surface finish: Black oxide

   Serviceable assembly loaded to 75% of fastener yield strength.

   Torque: _____ [N-m]

3. Fastener: M8 × 1.25

   Class: 8.8

   Stress area: 36.6 mm²

   Surface finish: Dry zinc-plated steel

   Serviceable assembly loaded to 75% of fastener yield strength.

   Torque: _____ [N-m]

Name _____   Date _____

Determine the power unit pressure setting to provide the required torque output for each bolting example using the following torque-wrench performance chart. *Note:* Pressure-torque performance charts are determined for each bolting tool by the manufacturer or calibration lab. Never use a pressure-torque performance chart unless it has been supplied for a specific tool. This is the main reason manufacturers require torque-wrench measurement verification with a calibration stand for each tool setup.

Performance Chart
PLARAD MX-EC 45 TS
S/N _____

Bar	PSI	N-m	Ft-lbs
80	1,160	484	357
100	1,450	600	443
120	1,740	716	528
140	2,031	832	614
160	2,321	948	699
180	2,611	1,064	785
200	2,901	1,180	870
220	3,191	1,296	956
240	3,481	1,412	1,041
260	3,771	1,528	1,127
280	4,061	1,644	1,213
300	4,351	1,760	1,298
320	4,641	1,876	1,384
340	4,931	1,992	1,469
360	5,221	2,108	1,555
380	5,511	2,224	1,640
400	5,802	2,340	1,726
420	6,092	2,456	1,811
440	6,382	2,572	1,897
460	6,672	2,688	1,983
480	6,962	2,804	2,068
500	7,252	2,920	2,154
520	7,542	3,036	2,239
540	7,832	3,152	2,325
560	8,122	3,268	2,410

Reference only. Courtesy of Maschinefabrik Wagner GmbH & Co.KG.

100 Preventative Maintenance

Name _____   Date _____

Determine the nearest full increment value on the pressure gauge for each of the following examples.

1. Torque: 600 N-m
   Pressure [bar]: _____

2. Torque: 1980 ft-lbs
   Pressure [psi]: _____

3. Torque: 1640 N-m
   Pressure [psi]: _____

4. Torque: 2070 ft-lbs
   Pressure [bar]: _____

Determine the gauge pressure reading for the following examples.

1. Pressure reading: _____ [bar]

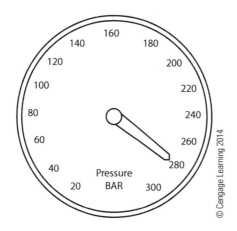

2. Pressure reading: _____ [psi]

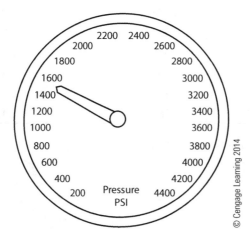

Worksheet 10–8

Name _____   Date _____

- List personal protective equipment (PPE) necessary to climb and work within a tube tower assembly with power-bolting equipment.

Hearing protection
_____
_____
_____
_____
_____
_____
_____

- _____
_____
_____
_____
_____
_____

-

CHAPTER 11

Machine Head

Name _____   Date _____

Use electrical schematics on the next three pages to determine the circuit protection devices necessary to isolate and LOTO the four yaw-drive motors, yaw-lubrication pump, and gearbox-lubrication pump.

Yaw-bearing lubrication-pump circuit protection (F): _____

Schematic page location: _____

Yaw motors 1 and 2 circuit protection (Q): _____

Schematic page location: _____

Yaw motors 3 and 4 circuit protection (Q): _____

Schematic page location: _____

Voltage level of yaw-motor drive circuits: _____

Gearbox lubrication-pump circuit protection (Q): _____

Schematic page location: _____

Worksheet 11–1

104 Preventative Maintenance

Name _____   Date _____

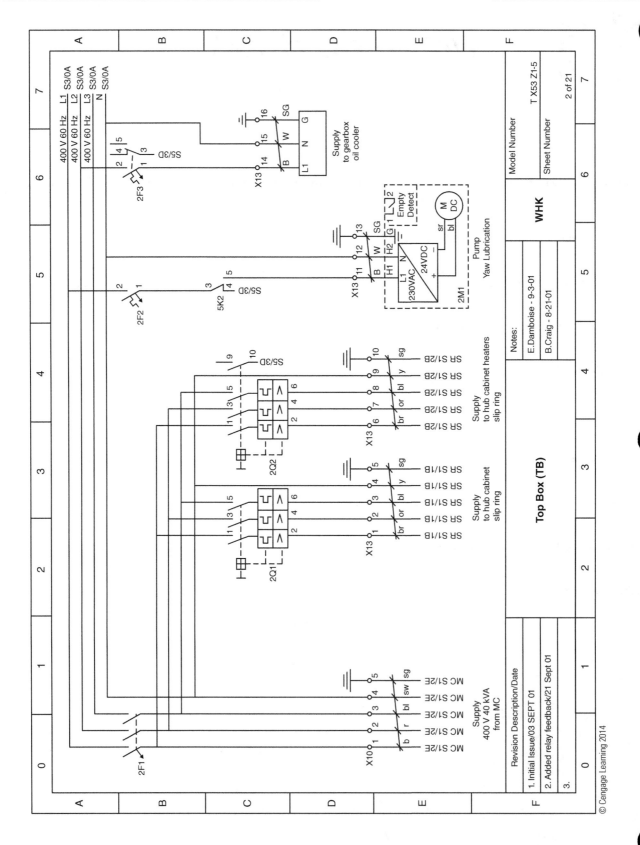

Worksheet 11-2

Name _____ Date _____

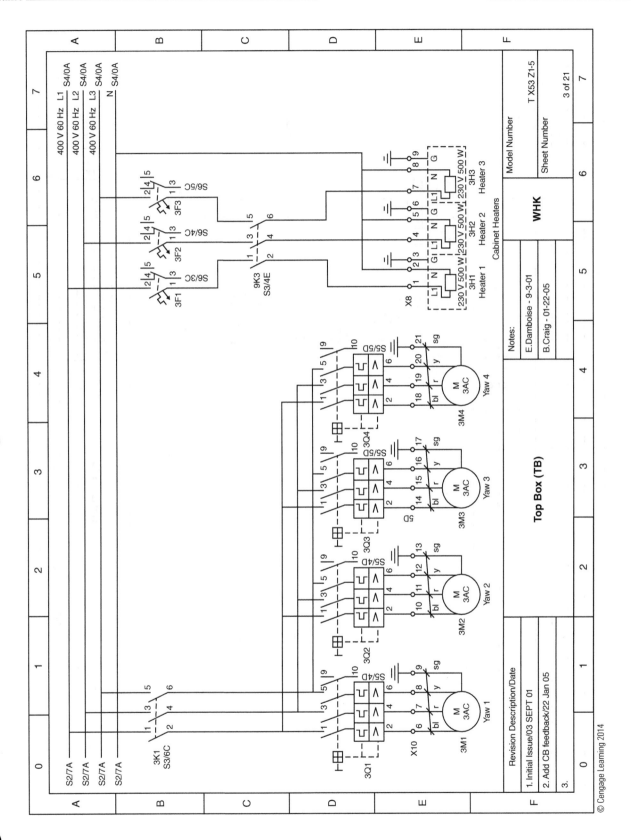

Worksheet 11–3

106 Preventative Maintenance

Name _____ Date _____

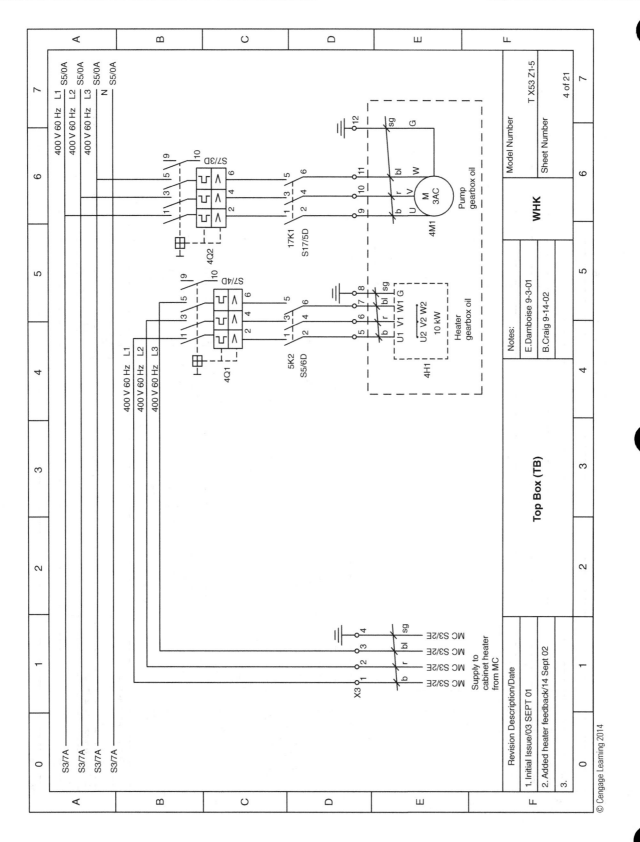

Worksheet 11–4

Chapter 11   Machine Head   107

Name _____   Date _____

● Name the electrical component shown at each location for the following schematic sheets.

1. Top box S2, 0E _____

2. Top box S2, 4C _____

3. Top box S2, 5E _____

4. Top box S3, 1E _____

5. Top box S3, 5B _____

6. Top box S3, 6E _____

7. Top box S4, 5D _____

8. Top box S4, 4E _____

9. Top box S4, 6E _____

● List the voltage level of the conductor(s) supplying the component listed on the following schematic.

1. Top box S2, 0E _____

2. Top box S2, 5E _____

3. Top box S3, 3E _____

4. Top box S3, 5E _____

5. Top box S4, 4E _____

6. Top box S4, 6E _____

●

Worksheet 11–5   Page 1 of 2

108   Preventative Maintenance

Name _____   Date _____

List the function of the electrical component shown at each location for the following schematic sheets.

1. Top box S2, 0E _____

2. Top box S2, 4C _____

3. Top box S2, 5E _____

4. Top box S3, 1E _____

5. Top box S3, 5B _____

6. Top box S3, 6E _____

7. Top box S4, 5D _____

8. Top box S4, 4E _____

9. Top box S4, 6E _____

Worksheet 11–6

Name _____   Date _____

● Identify the weld-joint type from the list for each of the following diagrams.

V-butt weld

Bevel-butt weld

Square-butt weld

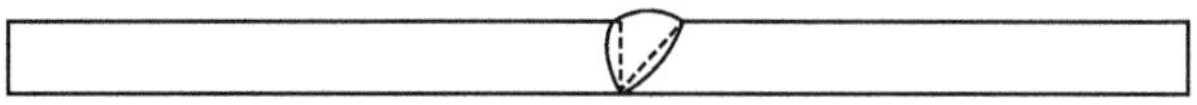

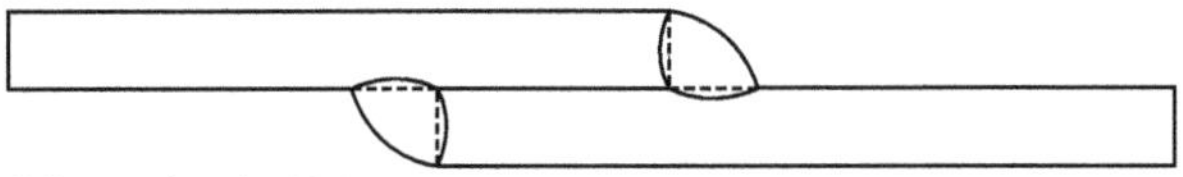

U-butt weld

Lap weld

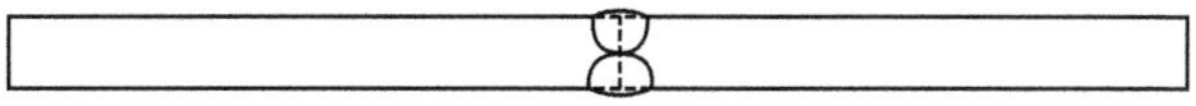

Fillet weld T-joint

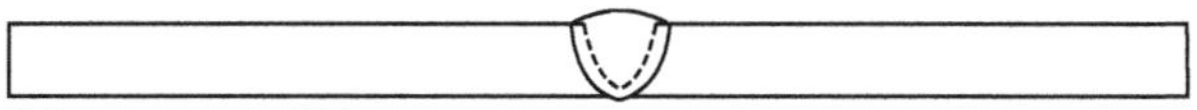

J-butt weld

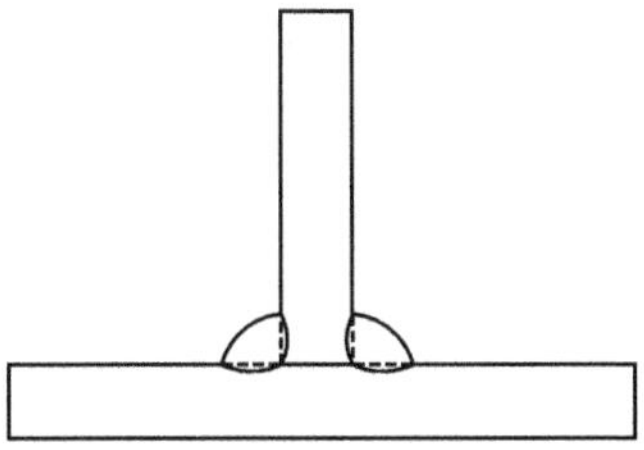

_____   _____

●

_____   _____

_____   _____

_____

●

Worksheet 11-7

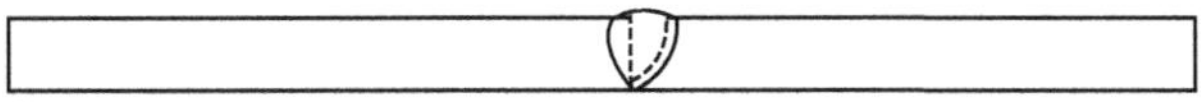

110   Preventative Maintenance

Name _____   Date _____

Complete the welding symbol callouts in the following table.

Weld Location	Butt-Weld Type				
	V	U	J	Bevel	Square
Arrow side		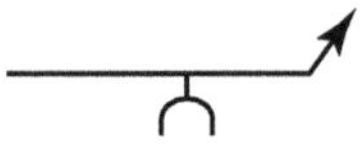	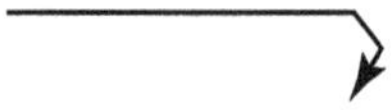		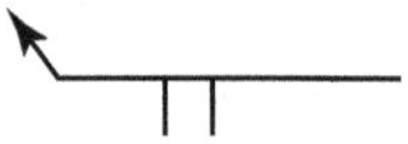
Opposite side		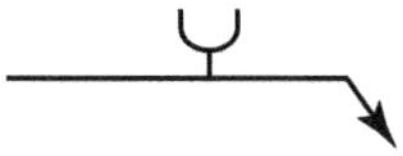			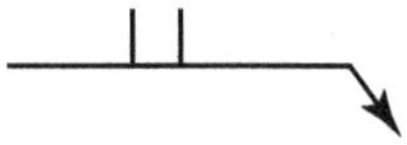
Both sides		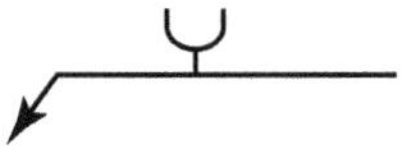	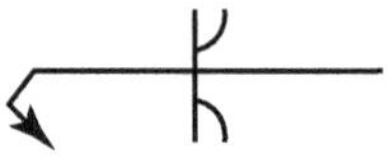		

Worksheet 11–8

Name _____   Date _____

Complete the schematic diagram.

Draw in the hydraulic valve assembly that would be electronically activated to operate the hydraulic yaw drive motors in stop, forward, and reverse.

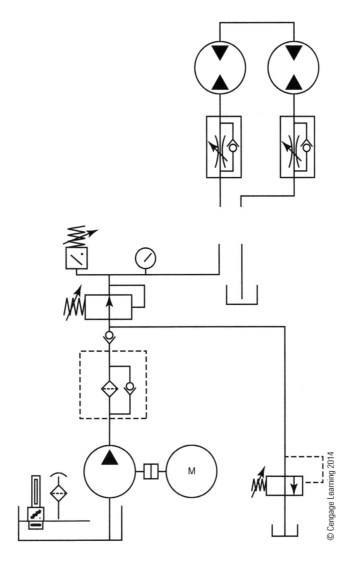

Worksheet 11-9

112   Preventative Maintenance

Name _____   Date _____

Review the electrical schematics on the next two pages for the PLC control circuit of the cooling damper assembly. Answer the following questions.

1. What is the designation of the digital input module for the damper limit switches? _____
   _____

2. What is the voltage supplied to the two limit switches? _____
   _____

3. What is the voltage supplied to the damper drive motor? _____
   _____

4. What is the voltage supplied to the PLC modules? _____
   _____

Name _____   Date _____

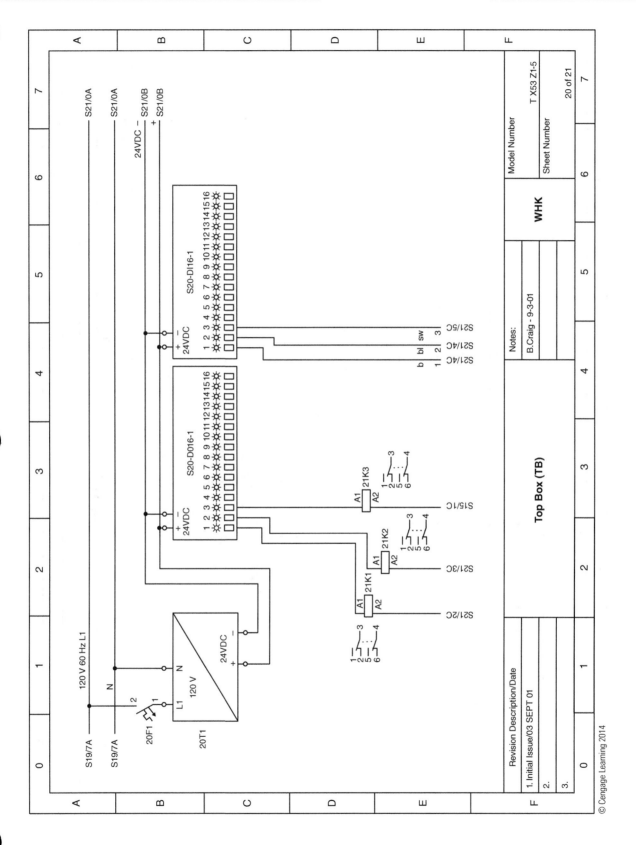

Worksheet 11–11

114 Preventative Maintenance

Name _____   Date _____

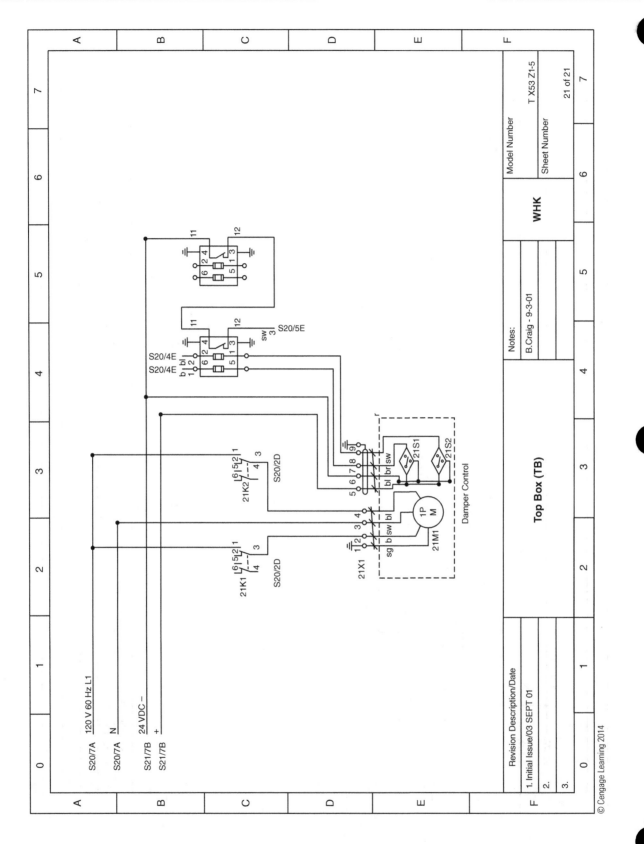

Worksheet 11–12   Page 3 of 3

CHAPTER 12

Drive Train

Name _____   Date _____

Use the following hydraulic schematic to determine the valve(s) necessary to isolate, relieve pressure, and LOTO the double-acting, spring-loaded cylinder of the parking-brake circuit and the four caliper-brake assemblies of the yaw-brake system.

Parking-Brake Assembly

Valve 1: _____

Schematic page location: _____

Valve 2: _____

Schematic page location: _____

Possible hydraulic pressure of the circuit: _____

Yaw-Brake Assembly

Valve 1: _____

Schematic page location: _____

Valve 2: _____

Schematic page location: _____

Possible hydraulic pressure of the circuit: _____

Worksheet 12–1

116 Preventative Maintenance

Name _____   Date _____

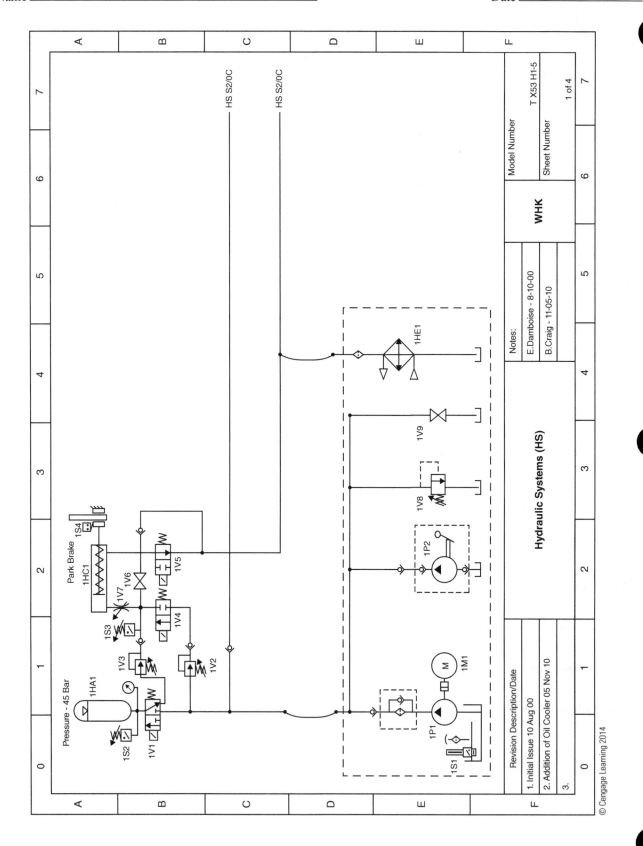

Worksheet 12–2

Name _____   Date _____

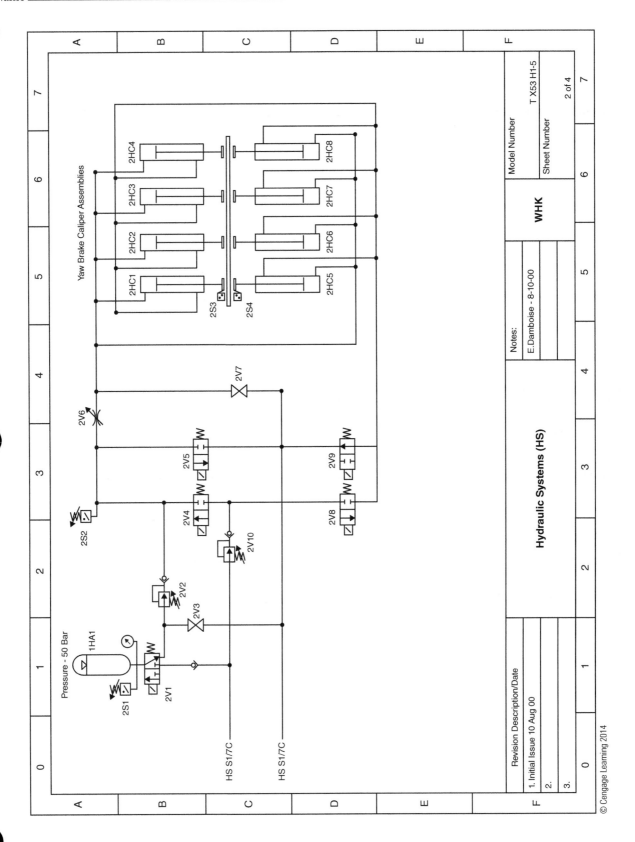

Worksheet 12–3

118 Preventative Maintenance

Name _____   Date _____

Name the hydraulic component shown at each location for the following schematic sheet.

1. Hydraulic systems S1, 2A _____

2. Hydraulic systems S1, 2B _____

3. Hydraulic systems S1, 1B _____

4. Hydraulic systems S2, 4C _____

5. Hydraulic systems S2, 1A _____

6. Hydraulic systems S2, 5B _____

List the hydraulic circuit function for each component determined above.

1. Hydraulic systems S1, 2A _____

2. Hydraulic systems S1, 2B _____

3. Hydraulic systems S1, 1B _____

4. Hydraulic systems S2, 4C _____

5. Hydraulic systems S2, 1A _____

6. Hydraulic systems S2, 5B _____

Worksheet 12–4

Chapter 12   Drive Train   119

Name _____   Date _____

Match the following drive-train components with an item shown on each drive-train diagram.

Oil-cooler assembly	Hub adapter	Main bearing
Compression coupling	Flex coupler	Main shaft
High speed shaft	Park brake	Brake disc
Brake caliper	Gearbox	Rotor lock
Lubrication pump	Pillow block	Air vent

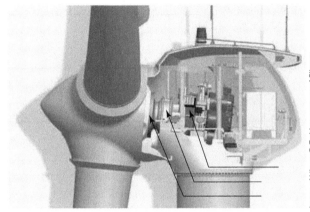

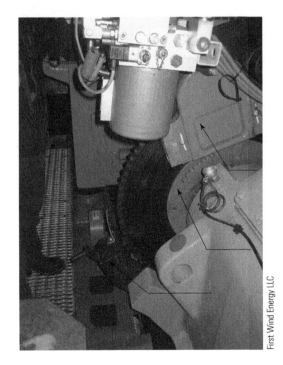

Worksheet 12–5

120 Preventative Maintenance

Name _____   Date _____

Oil Analysis

Match each oil element or contaminant with a possible source. There may be more than one source for an item, so use caution when reviewing an oil analysis and determining the possible source.

Iron	Lead	Chromium	Copper
Tin	Silver	Zinc	Silicon
Nickel	Organic	Phosphorus	Water

1. Roller bearing _____

2. Plain bearing _____

3. Gear _____

4. Shaft _____

5. Air vent _____

6. Lubrication pump _____

7. EP additive _____

8. Antifoam additive _____

Worksheet 12–6

Name _____   Date _____

Oil Analysis

Use the data provided to chart the following contaminant levels. Indicate on the trend chart which maintenance cycle a second oil analysis may need to be considered and the maintenance cycle when contamination levels are approaching the OEM's maximum specification.

Use Appendix E in the textbook as a reference for minimum and maximum acceptable levels.

Predictive Maintenance Data

Element (ppm)	Cycle						
	1	2	3	4	5	6	7
Iron	<1	<1	5	8	18	22	31
Copper	<1	<1	2	<1	8	<1	<1
Lead	<1	<1	<1	<1	1	1	<1
Zinc	370	375	366	378	384	402	396
Silicon	2	2	5	22	8	8	9
Water	59	59	72	101	68	70	74

122 Preventative Maintenance

Name _____ Date _____

Contaminant Trend - Iron

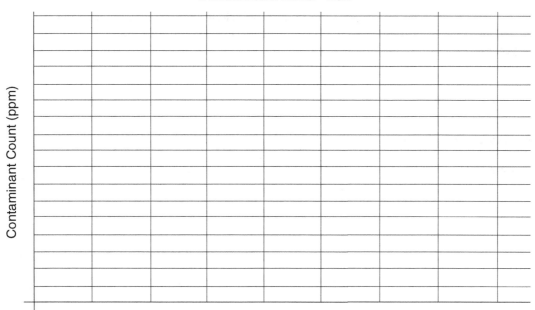

Preventative Maintenance Cycle

Contaminant Trend - Copper

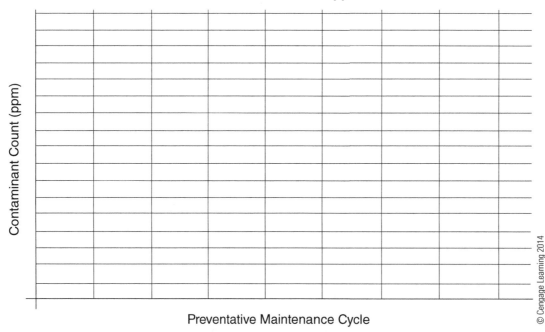

Preventative Maintenance Cycle

Name _____   Date _____

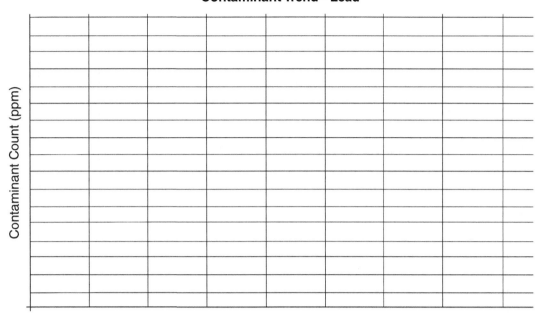

Contaminant Trend - Lead

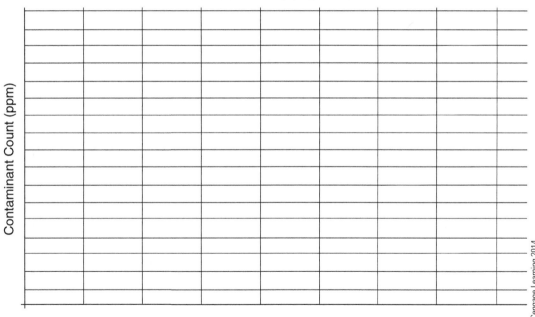

Contaminant Trend - Zinc

124 Preventative Maintenance

Name _____ Date _____

Contaminant Trend - Silicon

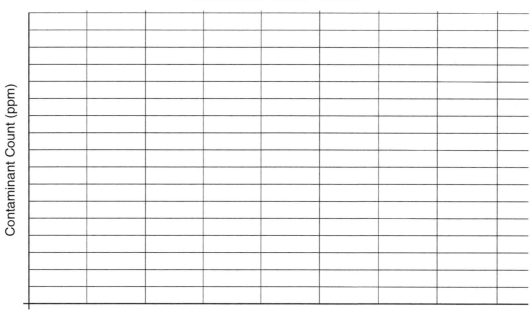

Preventative Maintenance Cycle

Contaminant Trend - Water

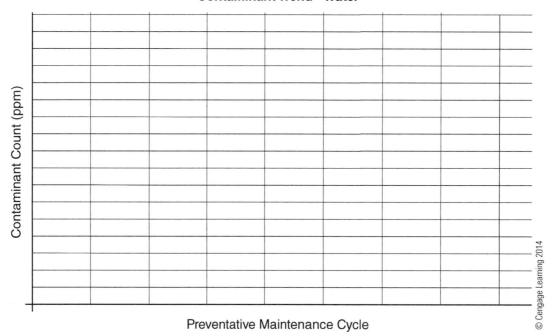

Preventative Maintenance Cycle

Name _____ Date _____

Answer the following MSDS questions on working with lubricating grease using Appendix I of the textbook.

1. Waste product number: _____

2. Date of issue for this MSDS: _____

3. Product trade name: _____

4. Product manufacturer: _____

5. First-aid measure for inhalation: _____

6. First-aid measure for ingestion: _____

7. Health hazards: _____

8. Fire-extinguishing media: _____

9. PPE for eye exposure: _____

10. Reactivity conditions to avoid: _____

CHAPTER 13

Generator

Name _____   Date _____

Match the following terms with the tool or component used with the generator alignment process.

Laser heads   Measuring tape   Dial indicator
Alignment control module   Jaw coupling   Magnetic base
Hydraulic power pack   Hydraulic cylinder   Hydraulic jack
Chain vise   Hydraulic wrench   Shim assortment
Manual hydraulic pump

1.

2.

3.

4.

5.

6.
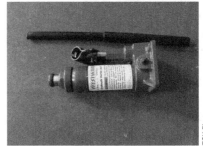

Worksheet 13–1

128 Preventative Maintenance

Name _____   Date _____

7.

8.

9.

10.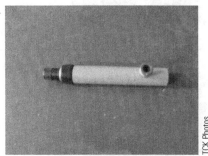

11.

12.

13.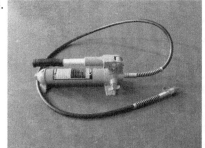

Worksheet 13–2

Name _____   Date _____

Determine an acceptable alignment tolerance for the following equipment parameters using the suggested maximum shaft misalignment values table.

Suggested Maximum Shaft Misalignment Values

Shaft Speed	Angular	Parallel	Angular	Parallel
RPM	mm/100 mm*	mm	0.001"/1 in.**	0.001 in.
0–1,000	0.10	0.13	1.0	5.1
1,000–2,000	0.08	0.10	0.8	3.9
2,000–3,000	0.07	0.07	0.7	2.8
3,000–4,000	0.06	0.05	0.6	2.0
4,000–6,000	0.05	0.03	0.5	1.2

*Gap at the coupling edge per 100-mm coupling diameter

**Gap at the coupling edge per 1" coupling diameter

1. Equipment operating speed 600 RPM and shaft coupling diameter 10":

   RPM range: 0–1,000

   Angular gap: 10" (coupling diameter) × 0.001"/1" = 0.01"

   Parallel offset: 0.001" × 5.1 = 0.0051"

2. Equipment operating speed 1,800 RPM, and shaft coupling diameter 560 mm:

   RPM range: _____

   Angular gap: _____

   Parallel offset: _____

3. Equipment operating speed 3,600 RPM, and shaft coupling diameter 3":

   RPM range: _____

   Angular gap: _____

   Parallel offset: _____

130 Preventative Maintenance

Name _____   Date _____

Determine the general operating condition for each of the following equipment setups given their vibration-measurement data compared to the sample vibration-severity chart provided. Issues that can create excessive vibration include shaft misalignment, unbalanced rotating assemblies, deformed rotating shafts, loose hardware, and damaged or dry bearings. Suggest a possible corrective action if the equipment setup has an unacceptable vibration level. There may be several possible corrective actions for any given situation depending on what equipment investigation reveals. Choose one issue listed above and suggest a possible corrective action.

Sample Machine Vibration Severity Chart

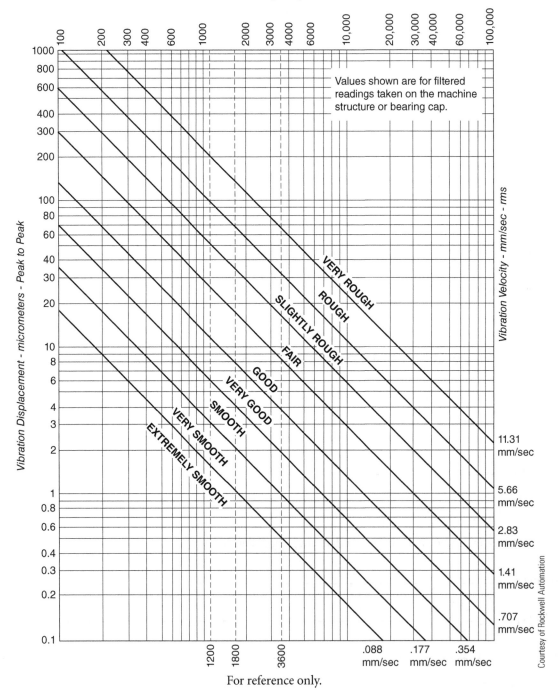

For reference only.

Worksheet 13–4   Page 1 of 3

Name _____   Date _____

1. Gear-motor assembly with gearbox output shaft bearing housing axial vibration measurement:

   Operating speed: 600 RPM

   Vibration displacement peak-to-peak reading: 80 micrometers

   Equipment condition: Fair

   Suggested corrective action: Inspect shaft bearing assembly for damage or lubrication level. Replenish lubrication if necessary and recheck vibration level.

2. Cooling fan assembly shaft end radial vibration measurement:

   Operating speed: 1,800 RPM

   Vibration displacement peak-to-peak reading: 115 micrometers

   Equipment condition: _____

   Suggested corrective action: _____

3. Generator rotor shaft end bearing housing axial vibration measurement:

   Operating speed: 3,600 RPM

   Vibration displacement peak-to-peak reading: 154 micrometers

   Equipment condition: _____

   Suggested corrective action: _____

Worksheet 13–5

132 Preventative Maintenance

Name _____ Date _____

4. Generator cooling-assembly vibration measurement appears to be elevated.

   Inspection of the cooling fan assembly indicates it is clean and moving freely. Measurement observed on the fan housing when the equipment is running and fan is not energized is 7.5 mm/sec–rms. Another system component to consider would be the cooling pump that operates at 1,800 RPM.

   Equipment condition: _____

   Suggested corrective action: _____

   _____

5. Cooling fan assembly shaft-end axial-vibration measurement:

   Operating speed: 1,800 RPM

   Vibration displacement peak-to-peak reading: 200 micrometers

   Equipment condition: _____

   Suggested corrective action: _____

   _____

6. Gearbox high-speed shaft-end bearing-housing radial-vibration measurement:

   Operating speed: 1,200 RPM

   Vibration velocity reading: 1.00 mm/second–rms

   Equipment condition: _____

   Suggested corrective action: _____

   _____

Worksheet 13–6

Name _____   Date _____

Use the following vibration-displacement data to fill in a predictive-maintenance run chart for a motor assembly operating at 1,800 RPM (use blank chart on next page). Measurement sample locations for each test are shaft end positions A and E. See the following diagram for details. Determine if corrective action is necessary for any of the maintenance cycles after reviewing the completed chart. For further practice, develop this graph with a spreadsheet software package.

Displacement Direction	Cycle 1	Cycle 2	Cycle 3	Cycle 4	Cycle 5	Cycle 6	Cycle 7	Cycle 8
Axial–E	0.2	0.4	0.5	0.65	1.5	3.5	8.0	15.5
Radial–A	0.25	0.5	1.1	8.2	30.6	71.5	103.5	212.8

Vibration displacement: mils, peak to peak.

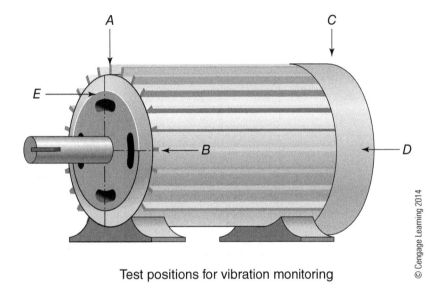

Test positions for vibration monitoring

Worksheet 13–7

134 Preventative Maintenance

Name _____ Date _____

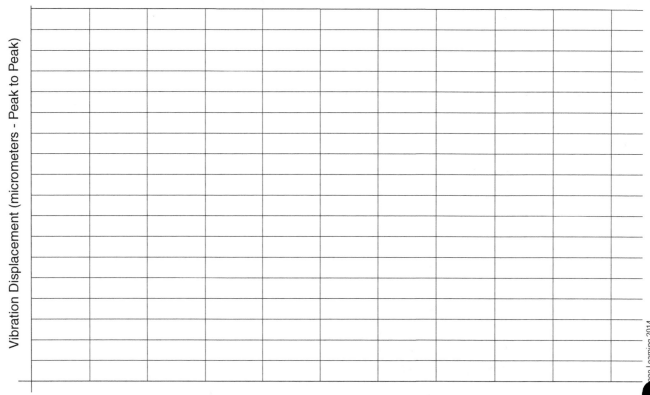

Was corrective action necessary? When? _____

_____

_____

Worksheet 13–8

Name _____   Date _____

Use the following insulation-resistance data to fill in a predictive maintenance log chart (next page) for a generator assembly. Measurements for each test are taken between the stator phase windings and ground. Determine if corrective action is necessary for any of the maintenance cycles after reviewing the completed chart. For this exercise, use minimum recommended winding insulation resistance of 100 kΩ. For further practice, develop a logarithmic graph with a spreadsheet software package.

Test Circuit	Cycle 1	Cycle 2	Cycle 3	Cycle 4	Cycle 5	Cycle 6	Cycle 7	Cycle 8
L1–GND	>2	>2	0.85	0.75	0.35	0.075	0.025	0.001
L2–GND	>2	>2	>2	>2	>2	1.85	1.80	1.75
L3–GND	>2	>2	>2	>2	1.65	1.62	1.55	1.55

Chart values are in megaohms. Winding insulation resistance is measured between phase and chassis ground cable (chassis isolated from ground). Always consult the test equipment manufacturer's recommendations for adjusting insulation resistance values for ambient temperature and humidity levels.

Worksheet 13–9

Name _____ Date _____

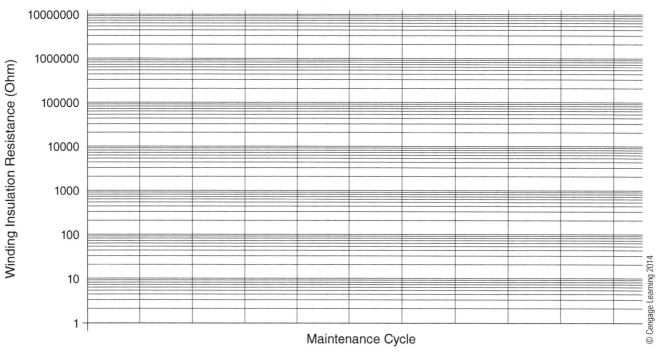

Was corrective action necessary? When? _____

_____

_____

Worksheet 13-10

CHAPTER 14

Rotor Assembly

Name _____   Date _____

Use electrical schematics on the next three pages to determine the circuit breaker(s) necessary to isolate and LOTO the hub main control cabinet heater and lighting circuits.

Hub main cabinet heater-circuit protection (Q): _____

Schematic page location: _____

Voltage level of heater circuit: _____

Hub lighting-circuit protection (F): _____

Schematic page location: _____

Voltage level of lighting circuit: _____

Worksheet 14–1

138 Preventative Maintenance

Name _____ Date _____

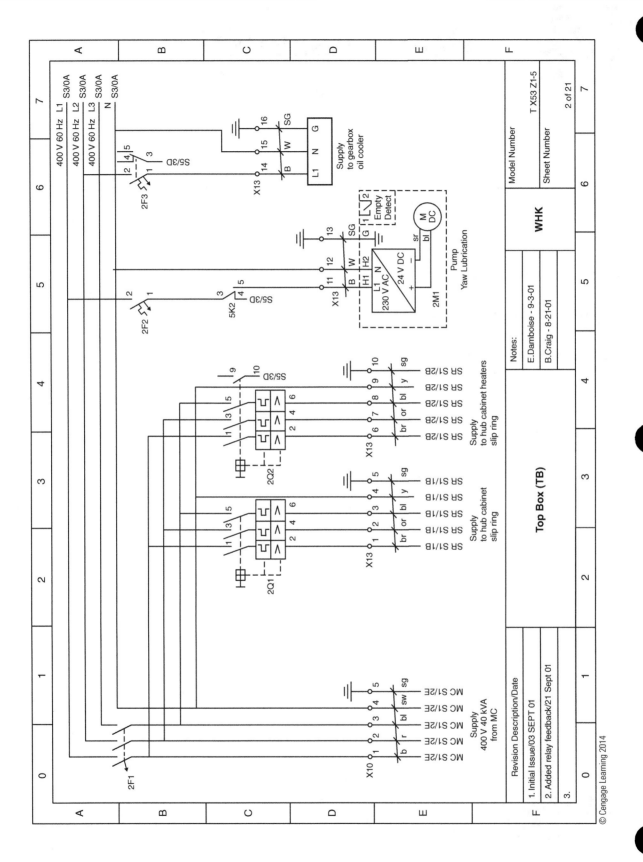

Worksheet 14–2

Name _____ Date _____

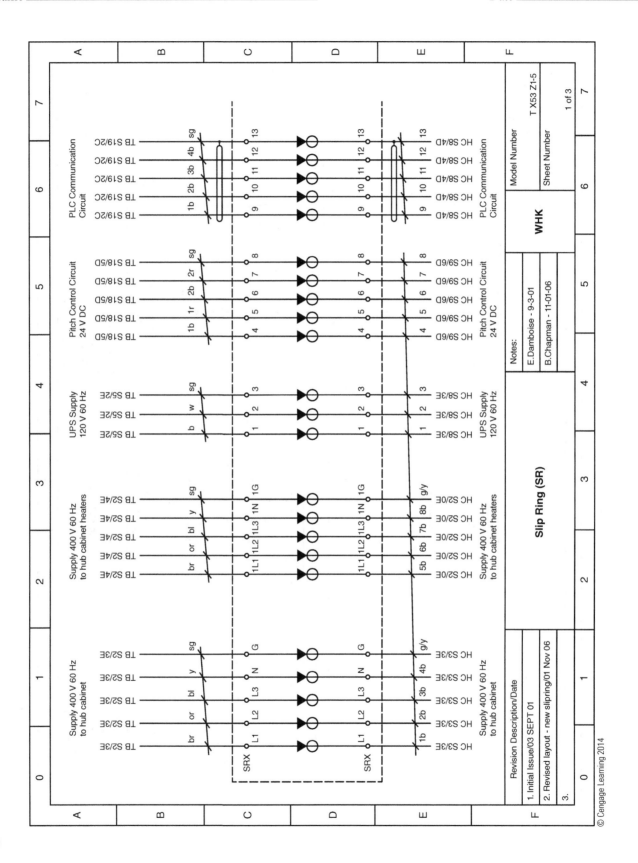

Worksheet 14–3

140 Preventative Maintenance

Name _____   Date _____

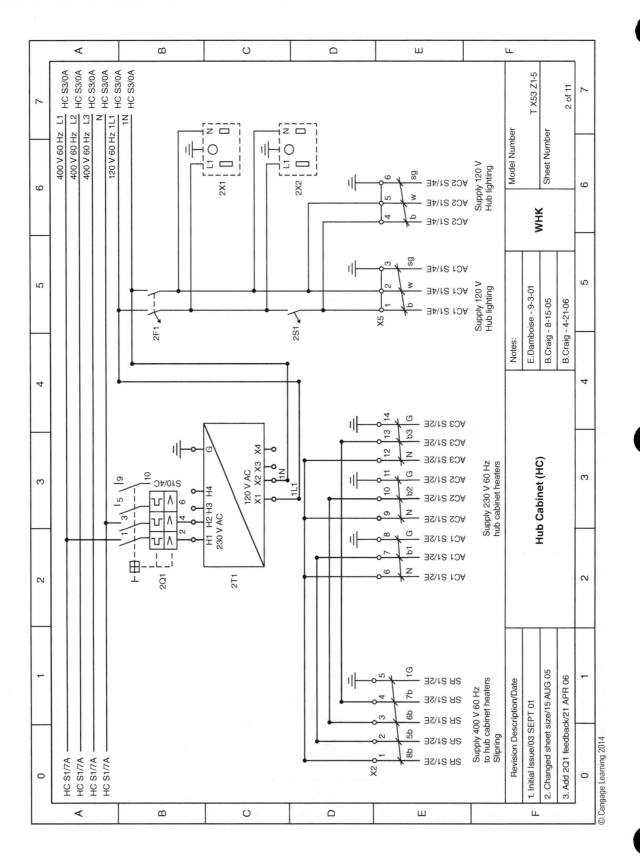

Worksheet 14–4   Page 4 of 4

Name _____   Date _____

Determine the applied tension (kN) for each fastener assembly and the power unit pressure (bar) to produce the tension. Typically, the fastener tension requirements are provided in the form of a work instruction or engineering specification for each assembly. In that case, it would be a matter of picking the pressure off the provided performance chart and adjusting the pump. This exercise requires practicing math similar to Chapter 10 of the workbook.

Tool Performance Chart

S/N ------------

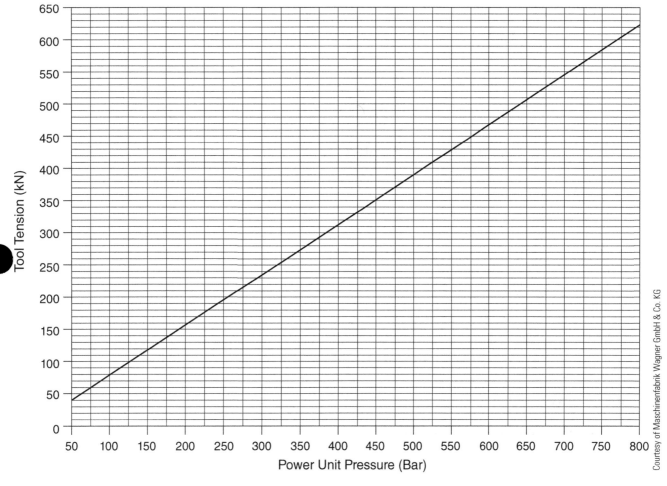

Worksheet 14–5

Name _____ Date _____

1. Fastener: M22 × 2.5

   Class: 10.9

   Stress area: 303.4 mm²

   Permanent assembly loaded to 90% of fastener yield strength.

   Applied tension: _____246_____ (kN)

   Power unit pressure: _____310_____ (bar)

2. Fastener: M30 × 3.5

   Class: 12.9

   Stress area: 560.5 mm²

   Serviceable assembly loaded to 75% of fastener yield strength.

   Applied tension: _____ (kN)

   Power unit pressure: _____ (bar)

3. Fastener M30 × 3.5

   Class: 10.9

   Stress area: 560.5 mm²

   Permanent assembly loaded to 90% of fastener yield strength.

   Applied tension: _____ (kN)

   Power unit pressure: _____ (bar)

4. Fastener: M27 × 3.00

   Class: 8.8

   Stress area: 459.4 mm²

   Serviceable assembly loaded to 75% of fastener yield strength.

   Applied tension: _____ (kN)

   Power unit pressure: _____ (bar)

Worksheet 14–6

Name _____ Date _____

- List some of the steps necessary to prepare for an external hub entry activity.

  Notify farm management of the entry activity start and estimated duration time _____
  _____
  _____
  _____
  _____
  _____
  _____
  _____
  _____
  _____

- _____
  _____
  _____
  _____
  _____

-

144 Preventative Maintenance

Name _____ Date _____

Fill in the hub entry steps of the following JSA for an external hub entry procedure. Use information gathered on sheet one of this exercise to fill in the description of activity portion of the form.

Job Safety Analysis

Tailgate Meeting Date _____

Wind Farm Name _____

Step	Description of Activity	Possible Injury	Corrective Action	Y/N
1				
2				
3				
4				
5				
6				
7				
8				
9				
10				
11				
12				
13				
14				
15				
16				
17				
18				
19				
20				

Wind Turbine Manufacturer & Model _____

Team Leader _____

Attendance Record _____

Safety Coordinator Review _____

Worksheet 14–8

CHAPTER 15

External Surfaces

Name _____ Date _____

List several external surface inspections for each of the following wind-turbine major components. Why would each inspection be important to the system's overall longevity?

Hub

_____
_____
_____
_____
_____

Blades

_____
_____
_____
_____
_____

Worksheet 15–1

Name _____   Date _____

Nacelle and Hardware

_____
_____
_____
_____
_____
_____
_____

Tower

_____
_____
_____
_____
_____
_____
_____

Chapter 15　External Surfaces

Name _____　　Date _____

● List necessary steps to safely exit a wind-turbine nacelle and inspect rotor blades and hub assembly.

Review the latest weather forecast for duration of the activity.

_____
_____
_____
_____
_____
_____
_____
_____
_____
_____

● _____

●

Worksheet 15–3

148 Preventative Maintenance

Name _____ Date _____

List each of the PPE, personnel requirements, and training requirements necessary to work safely on a wind-turbine nacelle.

PPE

Full-body harness
_____
_____
_____
_____
_____
_____

Personnel Requirements

_____
_____
_____
_____
_____

Training Requirements

_____
_____
_____
_____
_____

Worksheet 15–4

Name _____ Date _____

Complete the following JSA with activities, possible injuries, and corrective actions necessary to safely inspect rotor blades from the top of the nacelle assembly. Use information gathered on the first sheet of this exercise to assist with the process.

Job Safety Analysis

Tailgate Meeting Date _____

Wind Farm Name _____

Step	Description of Activity	Possible Injury	Corrective Action	Y/N
1				
2				
3				
4				
5				
6				
7				
8				
9				
10				
11				
12				
13				
14				
15				
16				
17				
18				
19				
20				

Wind Turbine Manufacturer & Model _____

Team Leader _____

Attendance Record _____

Safety Coordinator Review _____

CHAPTER 16

Developing a Preventative Maintenance Program

Name _____   Date _____

Review the bill of materials (BOM) on page 2 of this exercise and answer the following questions.

1. What is the assembly part number on the BOM? _____

2. What is the initial date of release for the BOM? _____

3. How many sheets are included with this BOM? _____

4. What is the revision letter for this BOM? _____

5. What is the part number of the 50-μ oil-filter element? _____

6. How many oil filters are required for the assembly? _____

7. What is the recommended grade of lubricating oil? _____

8. What is the oil-canister O-ring part number? _____

9. How many oil-canister O-rings are required? _____

10. What is the air-vent part number? _____

11. What is the blower-fan motor part number? _____

12. What is the voltage listing for the blower-fan motor? _____

13. What is the blower-blade part number? _____

14. What is the lubricating-pump part number? _____

15. What is the voltage listing for the lubricating-pump motor? _____

16. What is the lubricating-pump seal type? _____

Worksheet 16–1

Wind-Farm Management

Name _____   Date _____

Generic Drive-Train Gearbox Assembly

Sample Bill of Material

Bill of Material: 56-2658-STG Drive Assembly: LVB2021-C			Sheet Rev C: 5/22/10 Initial Issue 7/02/08
Item	Quantity	Description	Part Number/Rev
28	1	Gearbox, GR 70:1, 2.1 MW system type	56-2658-G70198/A
29	2	Pillow block, isolation type, neoprene, 80A Duro	75-3578-80D/L
30	1	Coupling, compression, 800 mm comp-type	88-7985-8/C
31	26	Bolt, hex head, M24-3 × 250 mm, class 10.9, BO	---
32	1	Frame, blower, XT-5155, galvanized	56-3335-G12/II
33	1	Blade, blower assembly, 350 mm, 5 B	76-5774-5B/D
34	1	Motor, blower, 3P, 400VAC, 50 Hz, face mt	55-3324-3P/C
35	1	Shroud, r-vinyl green, 30-45-600	22-344-600/D
36	1	Exchanger, A/L, 750-750-300	33-5678-E/C
37	12	Screw, SHC, M10-1.5 × 25 mm, class 8.8, BO	---
38	a/n	Hose, reinforced, 20 Bar, ID-52 mm	HTL-467/E, or equiv.
39	a/n	Hose, reinforced, 20 Bar, ID-32 mm	HTL-224/S, or equiv.
40	8	Clamp, adjustable, screw-type, stainless steel	22-3340ST/F
41	1	Frame, Lub. pump, RR-675, blue-epoxy coated	56-3345-E02/F
42	1	Pump, Lub, GT, 15LPM	445-8897/E
43	1	Motor, Lub. pump, 3P, 400VAC, 50 Hz, face mt	55-3321-3P/F
44	1	Valve, shut-off, swivel coupling	12-2247-SC/C
45	1	Canister, double-bell, 10 L	234-15-DB/C
46	1	Filter element, 50 μ-Single stage-FG-ST	HYDAC1510/-
47	1	Filter element, 10 μ/3 μ –Dual stage-FG	HYDAC16103/-
48	2	O-ring, canister, nitrile, 51 mm × 0.25 mm	Parker 35567/-
49	1	Seal, pump, V-pack, 30 D, spring, wiper	Parker 11267/-
50	270 L	Oil, lubricating, CGTX, 90 wt, synthetic	---
51	1	Sensor, pressure, 0–25 bar	GE67025B/B
52	1	Sensor, PT100, 8 mm-PT	Ω5763-8PT/C
53	1	Brake, park, Svenborg 5576-4	SVNB-5576/G
54	1	Vent, air, drum-type, desiccant, yellow	HYDAC33467/-
Notes: 1. a/n – "as needed" 2. Refer to individual part drawings for further details			Geartrain SP, LLC South Fork, IN Sheet 3 of 12
Eng. Approval: ___George Fournier, PE___ 5/22/2010 Mfg. Approval: ___Daniel Bevins, DMTS___ 5/07/2010			

Worksheet 16–2

Chapter 16   Developing a Preventative Maintenance Program   153

Name _____   Date _____

Match each of the following wind-turbine system components with the correct image on the following sheets.

Air vent (desiccant)	Yaw-drive lubrication	Oil-sight glass
Yaw-bearing grease manifold	Yaw-brake caliper	Main bearing
Generator mounting foot	Park-brake assembly	Gearbox lubrication pump
Gearbox pillow block	Gear-oil heat exchanger	Air-vent cap
Hydraulic power unit	Pitch-gear motor assembly	

Worksheet 16–3

154 Wind-Farm Management

Name _____   Date _____

Chapter 16 Developing a Preventative Maintenance Program 155

Name _____ Date _____

Worksheet 16–5

156 Wind-Farm Management

Name _____ Date _____

Use the information provided with the following component examples to complete an Internet search for manufacturer, model, size, and type information that may be used to cross reference the component or purchase a replacement.

Part Number(s): X30205M, Y30205M

Manufacturer: _____

Size: _____

Type: _____

Part Number(s): A4050, A4138D

Manufacturer: _____

Size: _____

Type: _____

Name _____ Date _____

Part Number(s): CR11138

Manufacturer: _____

Size: _____

Type: _____

Part Number(s): PS3866

Manufacturer: _____

Size: _____

Type: _____

158 Wind-Farm Management

Name _____ Date _____

Part Number: SNR 2206

Manufacturer: _____

Size: _____

Type: _____

Part Number: 5K42HN4127

Manufacturer: General Electric

Size: _____

Type: _____

Voltage: _____

Worksheet 16–8

Chapter 16 Developing a Preventative Maintenance Program 159

Name _____ Date _____

Part Number: CB101612

Manufacturer: _____

Size: _____

Type: _____

Part Number: 715-20-J

Manufacturer: Boston Gear

Size: _____

Type: _____

Worksheet 16–9

Wind-Farm Management

Name _____ Date _____

Determine some of the preventative and predictive maintenance items that may be used with the following hydraulic power units. Consider items highlighted in the images along with others that may not be pointed out such as wire connections, sensors, and hydraulic fittings. Use the following two sections to list maintenance items that should be considered on a semiannual and annual basis.

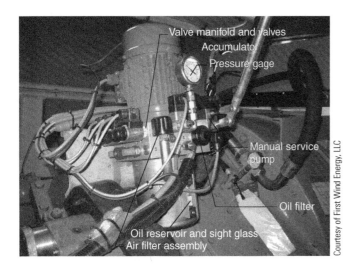

Preventative Maintenance Items

Semiannual _____

_____

_____

_____

_____

Annual _____

_____

_____

_____

_____

Worksheet 16–10

Name _____   Date _____

Predictive Maintenance Items

CHAPTER 17

Wind-Farm Management Tools

Name _____   Date _____

Place the correct term beside each of the following items. Indicate whether the item would be considered variable or attribute data.

11,345 kW   _____

Red   _____

603.5 VAC   _____

No   _____

Low voltage   _____

36,210 kWh   _____

Faulted   _____

On   _____

Running   _____

500 psi   _____

32 hours   _____

10 months   _____

1,500 amps   _____

3.5 quarts   _____

Hot   _____

Worksheet 17-1

Name _____   Date _____

Use the data provided to develop a line graph using spreadsheet software. The data shows the tool-performance relationship between torque and pressure for a power torque wrench. Develop a line graph for the 1MXT-SA torque output versus power unit pressure.

Power Torque Wrench
Performance Chart

Pressure (PSI)	Torque (Ft-Lbs)		Pressure (PSI)	Torque (Ft-Lbs)	
	1MXT-SA	3MXT-SA		1MXT-SA	3MXT-SA
1400	186	456	5800	778	1866
1600	214	512	6000	805	1930
1800	242	576	6200	832	1994
2000	270	640	6400	859	2058
2200	297	704	6600	886	2122
2400	324	768	6800	913	2186
2600	351	832	7000	940	2250
2800	378	896	7200	967	2316
3000	405	960	7400	994	2382
3200	431	1024	7600	1021	2448
3400	457	1088	7800	1048	2514
3600	483	1152	8000	1075	2580
3800	509	1216	8200	1101	2646
4000	535	1280	8400	1127	2712
4200	562	1346	8600	1153	2778
4400	589	1412	8800	1179	2844
4600	616	1478	9000	1205	2910
4800	643	1544	9200	1232	2974
5000	670	1610	9400	1259	3038
5200	697	1674	9600	1286	3102
5400	724	1738	9800	1313	3166
5600	751	1802	10000	1340	3230

Reference only   Courtesy of HYTORC-Wind, LLC

Worksheet 17–2

Name _____ Date _____

Use the data provided on sheet 1 to develop a line graph using spreadsheet software. The data show the tool-performance relationship between torque and pressure for two types of power torque wrenches. Develop a line graph that includes the 1MXT-SA and 3MXT-SA torque output versus power unit pressure on the same plot.

Wind-Farm Management

Name _____   Date _____

Use the data provided to develop a line graph with spreadsheet software. The data show changes in average monthly wind speed over a 12-month period. Develop a run chart of the wind-speed data and include a trend line to show how the average monthly wind speed varies throughout the year.

Month	Wind Speed (m/s)
January	3.2
February	3.3
March	3.6
April	2.7
May	2.1
June	2
July	1.8
August	2.2
September	2.6
October	2.7
November	2.8
December	3.0

What is the annual average wind speed? _____

Does the trend line indicate an increase or decrease in average monthly wind speed? _____

Worksheet 17–4

Name _____ Date _____

Use the data provided to develop a bar graph using spreadsheet software. The data table shows information gathered from a wind-farm SCADA status log. Use the bar graph to show the faults by type and frequency of occurrence.

SCADA Status Fault Report
11 November 2010

Date	Time	Turbine	Status	Condition
11-Nov-10	0:01:23	6	On-line	O113
11-Nov-10	1:33:01	73	Fault	E277
11-Nov-10	1:34:01	73	Shut down	O114
11-Nov-10	1:34:36	73	Ready	O111
11-Nov-10	1:35:22	5	Fault	E337
11-Nov-10	1:35:37	5	Shut down	O114
11-Nov-10	2:11:16	56	Fault	E277
11-Nov-10	2:12:16	56	Shut down	O114
11-Nov-10	2:13:01	56	Ready	O111
11-Nov-10	2:45:24	73	Run-up	O112
11-Nov-10	2:45:26	10	Run-up	O112
11-Nov-10	2:50:32	73	On-line	O113
11-Nov-10	2:50:44	10	On-line	O113
11-Nov-10	3:55:12	56	Run-up	O112
11-Nov-10	3:59:23	33	Run-up	O112
11-Nov-10	4:00:15	56	On-line	O113
11-Nov-10	4:04:26	33	On-line	O113
11-Nov-10	5:34:22	12	Fault	E080
11-Nov-10	5:34:37	12	Shut down	O114
11-Nov-10	7:30:57	15	Stop/Reset	O110
11-Nov-10	7:33:34	15	Ready	O111
11-Nov-10	7:35:55	15	Maintenance	M111
11-Nov-10	7:56:01	12	Service	S111
11-Nov-10	8:10:20	5	Service	S111
11-Nov-10	12:10:23	5	Stop/Reset	O110
11-Nov-10	12:13:07	5	Ready	O111
11-Nov-10	12:23:16	5	Run-up	O112
11-Nov-10	12:28:32	5	On-line	O113
11-Nov-10	14:11:44	12	Stop/Reset	O110
11-Nov-10	14:18:17	12	Ready	O111

Status information (S), operation (O), and error (E) codes are listed for reference only. Codes and format will vary depending upon the SCADA system provider. Fault: E277–No wind speed, Fault: E337–Gearbox over temperature, and Fault: E080–Generator speed not plausible

Wind-Farm Management

Name _____ Date _____

Use the data provided to develop a Pareto chart using spreadsheet software. The data table shows a sample of information gathered from technician service-call findings during a six-month period. Use the data to develop a bar graph showing the top three issues by occurrence over the six months.

Service Report Findings
July–December 2011

Date	Technician	Turbine	Fault	Corrective Action
2-Jul-11	MK	6	Main bearing over temp.	Tighten wire connection
4-Jul-11	MH	17	Gearbox over temperature	Replace PT 100
11-Jul-11	MH	19	Blade 3 not functioning	Tighten wire connection
12-Jul-11	JF	27	Yaw time out	Replace rectifier
15-Jul-11	DJ	5	Gearbox low oil	Add gear oil
20-Jul-11	MJ	5	Generator B2 over temperature	Tighten wire connection
23-Jul-11	MJ	2	Gearbox low oil	Add gear oil
27-Jul-11	MH	5	Gearbox over temperature	Tighten wire connection
29-Jul-11	MK	16	No wind speed	Replace CNT Module
30-Jul-11	JB	4	Generator winding over temperature	Tighten wire connection
3-Aug-11	JB	22	Battery voltage low	Tighten wire connection
8-Aug-11	JB	19	Converter over temperature	Replace fan motor
10-Aug-11	TB	28	Loss of hub COM	Tighten wire connection
13-Aug-11	MH	33	Cooling damper closed	Tighten wire connection
17-Aug-11	DJ	35	Gearbox low oil	Add gear oil
26-Aug-11	TB	34	Converter over temperature	Replace fan motor
6-Sep-11	TB	3	Main bearing over temp.	Tighten wire connection
9-Sep-11	TB	9	Gearbox over temperature	Tighten wire connection
10-Sep-11	MH	10	Yaw time out	Tighten wire connection
11-Sep-11	MH	15	Circuit breaker hub	Tighten wire connection
21-Sep-11	DJ	17	Loss of hub COM	Replace OV module
24-Sep-11	DJ	17	Loss of hub COM	Tighten wire connection
26-Oct-11	JB	9	Gearbox low oil	Add gear oil
11-Nov-11	MK	25	Gearbox low oil	Add gear oil
19-Nov-11	JF	26	COM timeout	Tighten wire connection
24-Nov-11	JF	5	Main bearing over temp.	Tighten wire connection
10-Dec-11	TB	4	Gearbox low oil	Add gear oil
15-Dec-11	MK	31	Generator winding over temperature	Tighten wire connection
25-Dec-11	JF	11	Main bearing over temp.	Replace DI16
30-Dec-11	JF	16	Gearbox low oil	Add gear oil

Worksheet 17–6

Name _____   Date _____

Use the following data to develop a project chart with project-planning software. Use the planning process to show minimum project timing and resources necessary for the permitting, development, construction, and connection of a residential wind turbine. Data necessary for this chart include the following.

Activity	Duration (Days)
Order and receive turbine	30
Order and receive tower	30
Lead time for crane	10
Lead time for utility inspection	7
Permitting (community, DEP, FAA)	30
Tower assembly	2
Turbine assembly	2
Foundation site preparation	3
Foundation assembly	2
Lead time for concrete delivery	10
Concrete pour	1
Concrete cure time before tower assembly	30
Erection of turbine with crane	1
Cable trench, assembly, and burial	2
Electrical connections	1
Utility inspection	1

Step 1 Layout the process activities to take advantage of lead times. The goal is to use parallel activities to minimize overall project time whenever possible.

Step 2 Enter each activity in the project software in the order developed in step 1.

Step 3 Enter the duration of each activity and set prerequisites in the appropriate column to show the relationship between each of the activities.

Step 4 Adjust the schedule if any steps can be optimized to reduce the overall project timing.

Step 5 Print out the schedule data and chart on separate sheets to show the completed exercise. How is understanding project flow and lead times beneficial when developing project strategies?

Worksheet 17–7

APPENDIX A

Hydraulic Symbols

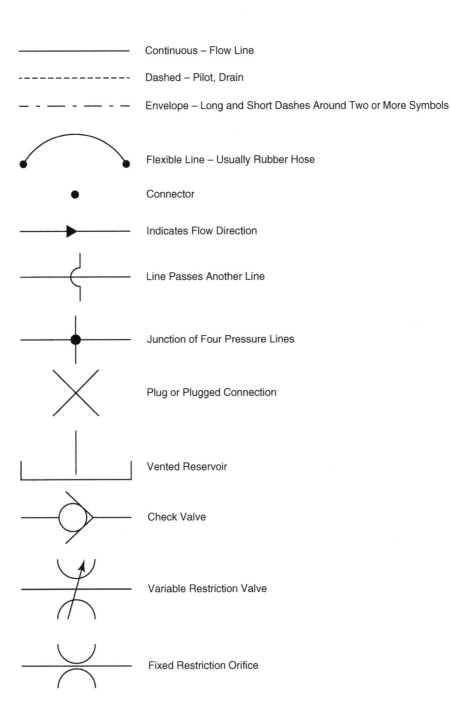

172　Appendix A　Hydraulic Symbols

 Quick Disconnect, without Check Valve

 Quick Disconnect, with Check Valve

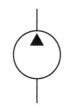 Fixed – Displacement Pump

 Variable – Displacement Pump

  Electric Motor

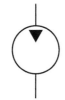 Fixed – Displacement Hydraulic Motor Single Direction

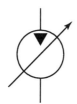 Variable – Displacement Hydraulic Motor Single Direction

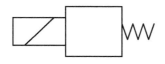 Solenoid Control Valve with Spring Return

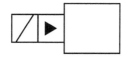 Solenoid Control Valve Operated by a Hydraulic Pilot

  Hydraulic Pilot Control Valve

Appendix A Hydraulic Symbols **173**

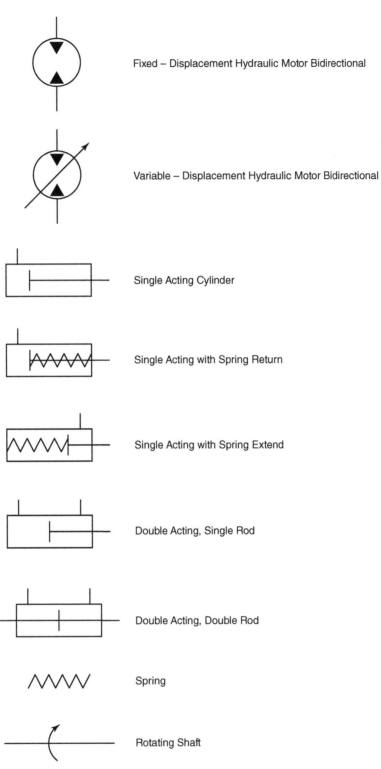

Appendix A Hydraulic Symbols

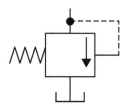 Pressure Relief Valve, Normally Closed

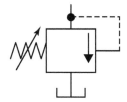 Variable Pressure Relief Valve, Normally Closed

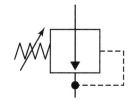 Variable Pressure Reducing Valve, Normally Open

 Metered Flow to Right and Free Flow to Left

 Directional Control Valve, Normally Closed, Two Position and Two Connections

 Directional Control Valve, Normally Open, Two Position and Two Connections

 Directional Control Valve, Two Positions, Four Connections

 Directional Control Valve, Two Positions, Three Connections, Normally Open

 Directional Control Valve, Two Positions, Three Connections, Normally Closed

Appendix A Hydraulic Symbols **175**

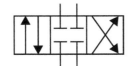 Directional Control Valve, Three Positions, Four Connections, Center Position Closes the Valve

 Directional Control Valve, Two Positions, Five Connections

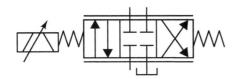 Proportional Direction Control Valve, Analog Electrical Signal Input Provides a Similar Analog Fluid Power Output (Electro – Hydraulic Servo Valve)

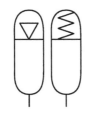 Accumulator, Gas Charged, and Spring Load

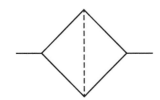 Filter or Strainer

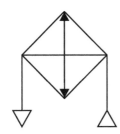 Heat Exchanger Air Cooled

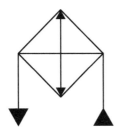 Heat Exchanger Liquid Cooled

176 Appendix A Hydraulic Symbols

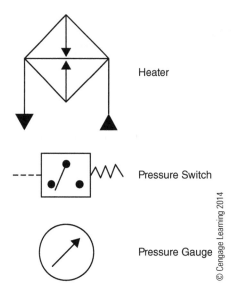

APPENDIX

Electrical Symbols

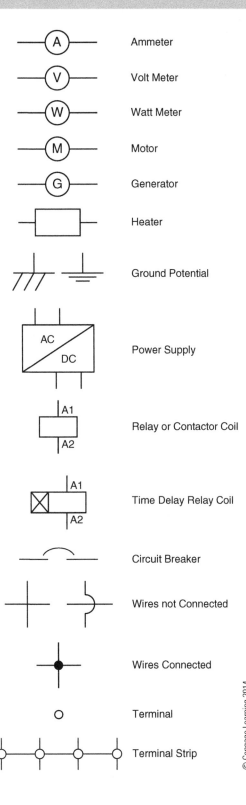

Appendix B   Electrical Symbols

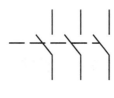 Multiple Switches with Mechanical Connection

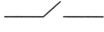 Switch

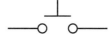 Push Button (NO)

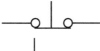 Push Button (NC)

 Relay or Contactor NO/NC Terminal Pair

 Shielded Conductors in a Cable

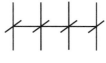 Multiple Conductors in a Cable

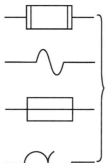 Fuse or Over Load Device

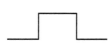 Indicates Thermal Overload Device

 Battery Cell

 Battery

 Capacitor (Fixed)

 Capacitor (Variable)

© Cengage Learning 2014

Appendix B Electrical Symbols

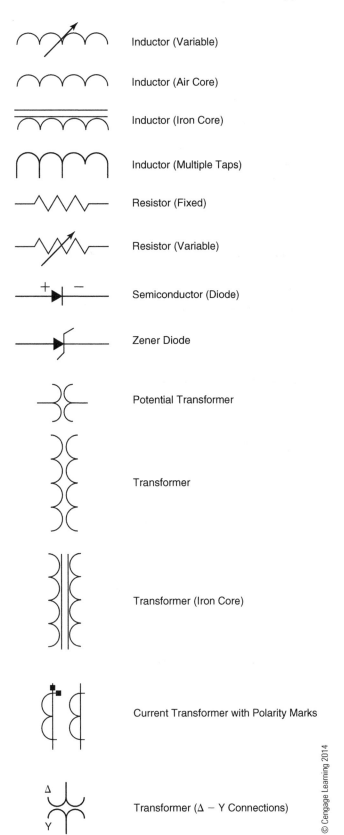

180 Appendix B Electrical Symbols

 Transformer (Δ – Δ Connections)

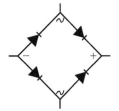 Rectifier (Full Bridge)

 Light – Emitting Diode (LED)

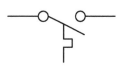 Temperature Switch (NO)

 Pressure Switch (NC)

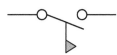 Flow Switch (NO)

 Liquid Level Switch (NC)

 Limit Switch (NO)

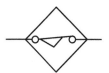 Proximity Switch (NC)

APPENDIX C

Bolting Notes

Bolt Torque Calculations

$$T \text{ (in.-lb}_f\text{)} = KD(\text{in.}) F(\text{lb}_f) \qquad T(\text{N-mm}) = KD(\text{mm})F(\text{N})$$

$$T(\text{N-m}) = [KD(\text{mm})F(\text{N})]/1{,}000 \qquad T \text{ (ft-lb}_f\text{)} = [KD(\text{in.})F(\text{lb}_f)]/12$$

T: torque

K: friction value (see the following chart)

D: nominal diameter of bolt

F: force applied to stretch bolt (bolt acts like an extension spring)

K Values for a Variety of Thread Surface Finishes

K	Surface
0.10	Waxed
0.15	Lubricated
0.18	Dry, zinc plated
0.20	Slightly oily, plain or black oxide
0.25	Hot dip galvanized

Sample Problem

Determine the torque (ft-lbs) required for a 1/2-13UNC to create a 10,000 lb_f tension with a hot dip galvanized finish.

K = 0.25

D = 0.50

F = 10,000 lb_f

T = [(0.25)(0.50 in.)(10,000 lb_f)]/12

T = **104 ft-lbs**

Threads

U.S. Standard (SAE) thread per inch (TPI): e.g., 1/2-13UNC

Nominal diameter of bolt is 1/2 or (0.50) inches.

TPI for the bolt is 13 threads per inch or 1/13" = 0.077" from thread peak to peak.

Metric standard (ISO) thread pitch: e.g., M20 × 1.25

Nominal diameter of bolt is 20 millimeters.

Thread pitch for the bolt is 1.25 mm, and this is also the measurement from the adjacent thread peaks.

Strength of Bolt Materials

Sample Problem

Determine the elongation of the 1/2-13UNC bolt if the length between the nut and the head is 3".

$\delta = Pl/AE$

$P = F = 10{,}000$ lbf

$l = 3"$

$A = 0.1419$ in.^{2} (from engineering standards book)

$E = 30{,}000{,}000$ psi (from engineering standards book)

$\delta = [(10{,}000\ \text{lb}_f)(3")]/(0.1419\ \text{in.}^2)(30{,}000{,}000\ \text{psi}) = \boxed{0.007"}$

Sample Problem

Determine the tensile force on a 3/8-16UNC threaded stud when the nut is turned one full revolution after the assembly is snug to the flange. Active length of the threaded rod is 10" between the threaded connection and the nut. Be sure to use the proper units to achieve the correct answer.

$\delta = Pl/AE$

$l = 10"$

$A = 0.0775$ in.^{2} (from engineering standards book)

Distance between thread peaks is $1"/16 = 0.0625"$ [elongation (δ) results from one full revolution of the nut]

$P = \delta AE/l = [0.0625"(0.0775\ \text{in.}^2)30{,}000{,}000\ \text{psi}]/10" = \boxed{14{,}531.5\ \text{lb}_f}$

Bolt Stress

Stress = F/A

Stress in MPa or PSI, depending on the system of units used.

F = force (N or lb$_f$)

A = area (in.^{2} or mm^2)

1 MPa = 1 N/mm^2

Use bolting engineering standards to determine the proper bolt material and temper according to the minimum yield strength value for each bolt grade or class.

APPENDIX D

Useful Information

Unit Prefixes

Prefix	Symbol	Multiplication Factor
tera	T	$1{,}000{,}000{,}000{,}000 = 10^{12}$
giga	G	$1{,}000{,}000{,}000 = 10^{9}$
mega	M	$1{,}000{,}000 = 10^{6}$
kilo	k	$1{,}000 = 10^{3}$
hecto*	h	$100 = 10^{2}$
deka*	da	$10 = 10^{1}$
deci*	d	$0.1 = 10^{-1}$
centi*	c	$0.01 = 10^{-2}$
milli	m	$0.001 = 10^{-3}$
micro	μ	$0.000\,001 = 10^{-6}$
nano	n	$0.000\,000\,001 = 10^{-9}$
pico	P	$0.000\,000\,000\,001 = 10^{-12}$

*Prefix and symbol are typically avoided with common naming practices.

Common Unit Conversions

U.S. Customary Unit	
Length	
1 foot	12 inches
1 mile	5,280 feet
1 yard	3 feet
Area	
1 square foot	144 square inches
9 square feet	1 square yard
1 acre	43,560 square feet
1 square mile	640 acres
1 circular mil	area of a circle 0.001 inch in diameter
1 circular inch	1,000,000 cir mils
Volume	
1 cubic foot	1,728 cubic inches
1 gallon	231 cubic inches
1 cubic yard	27 cubic feet

Liquid measure

1 cup	8 ounces
1 pint	16 ounces
1 quart	2 pints
1 gallon	4 quarts
1 cubic foot	7.4805 gallons

Weight

1 pound	16 ounces
1 short ton	2,000 pounds
1 long ton	2,240 pounds

Circular measures

1 minute	60 seconds
1 degree	60 minutes
1 quadrant	90 degrees
1 circumference	360 degrees
1 radian	57.2958 degrees

Unit Conversions

Length

1 inch	25.4 mm
1 inch	2.54 cm
1 foot	0.3048 m
1 yard	0.9144 m
1 meter	3.281 feet
1 kilometer	3,281 feet
1 mile	1,609 m

Area

1 square meter	1,550 square inches
1 square meter	10.76 square feet
1 square meter	1.196 square yards
1 square foot	0.0929 square meters
1 square yard	0.836 square meters
1 acre	4,047 square meters
1 square mile	2,589,988 square meters

Mass

1 pound	0.4536 kilograms
1 kilogram	2.205 pounds
1 short ton	0.907 metric tons
1 metric ton	1.102 short ton
1 long ton	1.016 metric tons
1 metric ton	0.984 long ton

Pressure

Pascal(N/m^2)	0.000145 psi
1 bar	14.504 psi

1 bar	100,000 Pascal
1 atm	101,326 Pascal
1 atm	14.696 psi
1 psi	6,894.8 Pascal

Velocity

1 m/s	2.237 mph
1 km/h	0.6214 mph
1 ft/s	0.3048 m/s
1 ft/s	1.097 km/h
1 mph	0.4470 m/s
1 mph	1.609 km/h

Acceleration

1 m/s^2	3.281 ft/s^2
9.8 m/s^2	32.15 ft/s^2

Power

1 Hp	0.7457 kw
1 kw	1.341 Hp
1 Hp	550 ft-lb$_f$/s
1 Hp	1 Btu/s

Temperature

0° Celsius	32 °Fahrenheit [t °C = (t °F − 32)/1.8]